ROTOL

The History of an Airscrew Company

1937–1960

Aerial view of Rotol's new factory in 1938

ROTOL

The History of an Airscrew Company

1937–1960

ISBN 0 9516815 0 8

Produced by Alan Sutton Publishing Ltd, Stroud, Glos.
Printed in Great Britain at Bookcraft Bath Ltd.

Contents

Author's Notes and Acknowledgements vii
Early History of Propellers 1
The Birth of Rotol Airscrews 4
Settling in at Staverton 13
The Early War Period 25
Dispersal to Gloucester and Cheltenham 40
The Mid-War Years 56
Electric Propellers and Further Changes 72
The Closing Months of the War 87
1945 – Dramatic Changes at Rotol 97
A Nautical Excursion 104
Sport and other Entertainment 111
Rotol at Peace 119
Into the 'Fifties 128
The Dowty Take-over 144
British Messier 152
Notes on Sources 161
Appendices 166

Dedication

To my wife June, without whose help and patience throughout the months of research and inevitable upheaval it caused, none of this would have been possible.

Author's Notes and Acknowledgements

A few weeks before I completed the final draft of this book, a profound change took place at Dowty Rotol when the name of the company was altered to Dowty Aerospace Gloucester and further changes in the structure of the company followed. As from June 1st 1990, five separate divisions were set-up, effectively re-establishing the original design and manufacturing of each product. Exactly how this will work out it is impossible to say at this stage. One can only hope that the future continues to remain bright for the company, although the challenge of foreign competition is greater today than at any time in its history. Rather than confuse the text of the book by altering references from Dowty Rotol to Dowty Aerospace I have chosen to leave the name Dowty Rotol, which was its official name from 1960 to 1990.

The present book began life when the author, as a recently retired employee, attended the Dowty Rotol Open Day in September 1988 and saw reference in the programme to the fact that the company were also celebrating fifty years on the Staverton site. Although there had been magazine articles in several periodicals over the years, no complete record of the company existed and it seemed a pity not to attempt to chronicle the Rotol story while many of the original employees were still able to contribute. The history of any company is inextricably entwined with the people who worked there and I was determined to ensure that this theme would form the background, interwoven with the history of the company itself.

With the blessing of the Dowty Rotol management, I began to contact various people and to circulate them with a brief questionnaire. However, I must make it clear at the outset, although written with considerable help from the present company, all the opinions expressed in its pages are my own. Apart from some minor alterations no form of censorship has been involved. This point should be borne in mind when certain parts may appear rather critical of the period, and it is much to the credit of the present company management that they have not sought to restrict my findings.

The book only covers the period from 1937 to 1960, not because I wished to exclude the progress which has been made by the company since that date, but from 1960 both Rotol and British Messier operated as part of the Dowty Group and the story of Dowty has already been adequately dealt with in two previous volumes.

Although the chapters in this book are laid out in a natural time sequence, starting with the early history of propellers, the creation of Rotol and progressing to the Dowty take-over, it has meant that certain items occur in more than one chapter, as the company evolved and new products came along. In order to save the reader dodging from one chapter to another the author has chosen to include British Messier, Nautical Matters, and Sports/Pastimes into single chapters which stand on their own, thus overcoming this basic defect.

In the following pages will be found a selection of personal anecdotes and incidents which were related to me in the course of the research for this book. Without these individual stories to enliven the more mundane procession of dates, events and products, the result might have been a dreary exercise and I trust that readers will not find, despite my best endeavours, that the result IS dreary. Anyone who remembers their days with the company will surely find something or someone they once knew in these pages and this will, I hope, bring back memories of happy times. Readers who did not have the good fortune to work at either Rotol or British Messier, may regard the book as a glimpse of life in 'the good old days' of a British engineering company, at a crucial period of the Twentieth Century. Some of the events of which I write happened more than fifty years ago, plenty of time for them to have acquired the title of 'The Good Old Days' which is how most of us remember our youth. It is interesting to speculate that the youngsters of today will almost certainly look back at the 1990's from their vantage point of the 2040's as being their 'good old days' and perhaps even write a book about their recollections.

Memories are notoriously fickle, enabling us to recall days long past, far more clearly than those of yesterday. Events which happened many years ago have possibly acquired an aura of legend, with the result that certain stories may not be exactly the same as the one that you recall. It has proved particularly difficult to precisely date all of the events and even more difficult to keep pace with promotions of individuals, as not much is retained in the company on this subject. If I have given anyone the wrong title, I trust I will not be castigated too severely. Wherever possible I have asked individuals to proof read parts of the manuscript, in an endeavour to check certain facts but in the last analysis the author accepts the blame for anything which is incorrect. If correction is needed, or anyone has further information which could be useful, I will be pleased to receive it for inclusion in any future edition.

To all those who helped me in the task of getting enough background stories, I give my thanks. The success of the book will be due in no small measure to their contributions and I have given their names below, but if you find that I have missed out on your own particular side of Rotol or British Messier life I offer my apologies. A few tales had to be excluded for a variety of reasons but almost everything that lies in these pages came from someone who wanted to make sure I didn't miss his or her story and I have tried to include almost all I was offered, even if it had to be slightly edited along the way.

Anyone who is looking for a recounting of the few scandalous goings-on will be disappointed and I have tried to avoid making any slanderous accusations against individuals! Most readers will be aware of the times when so-and-so was hauled up before the supervisor – or the dreaded personnel department if it was really serious – for transgressing the rules in some form or another. This sometimes stemmed from a little bit of private enterprise by the individual concerned, who perhaps needed to repair part of his motor car or lawn mower. There are many tales of parts being reground, painted, plated or completely refabricated and, if he was really unlucky, the unfortunate employee getting caught in the act. All these activities came under the general heading of 'foreigners' – which were frequently produced to far higher standards than the original and, should the need be urgent, in record time. Although I was told some very interesting tittle-tattle I decided it was kinder to concentrate on the main historical facts.

For much the same reason I have edited out all the stories concerning the occasional illicit romances, whether serious or passing. In the majority of cases they were almost certainly harmless and they happened a long time ago. One can also balance this with the knowledge that both Rotol and British Messier ran an unpaid but flourishing matrimonial agency, resulting in a great many marriages over the years and I believe that most have lasted the course.

Despite all my efforts, it is probably inevitable that something has been left out but I sincerely hope it is not too much and that it does not greatly affect the whole picture. There are several people, who played a significant part in the history of the company, that I should like to have consulted, but for one reason or another it was not possible. Time has already taken some from us and even during my research, others followed them – one in particular was Sir John Evetts who died at the age of 97, only a few days before I started the project. Another was Hugh Conway, whose passing was mourned by his colleagues at British Messier. Many of those who were with the company in the early days are now in their eighties and have pleaded that their memories are too hazy and have therefore asked to be excused. There are some notable exceptions, for which I am most grateful.

Although the memories of employees formed the background of this story, a significant amount came from newspaper articles and contemporary accounts, most of which are contained in magazines of the period. Nearly all of these are to be found in the Technical Library of the present company, but some are in private hands. I have also used the two house magazines, firstly *The Airscrew*, from the Second World War and in the mid 1950's the *Rotol and British Messier Quarterly Review*. Although every effort has been made to ensure the authenticity of all source material, it would be unlikely if complete accuracy could be claimed for every event which I have retold in these pages. In addition, I was fortunate to locate a few individuals who had hoarded some old documents, magazines, press clippings and the like, which would normally have been thrown away. To these 'squirrels' – of whom I was a prime example – I express my particular thanks!

A most important source of material is contained in the minute books of the meetings of the Rotol Board of Directors, which the present Secretary of the Dowty Group of Companies, Mr C.J.M. Walker, was most kind in locating and allowing me access. The value of these books cannot be overemphasised, as they enabled me to get a much clearer picture of certain decisions that were made and afforded an insight into the problems encountered by the Rotol Directors over the years. There are three books altogether; the first spans the years from May 1937 to December 1945 and the third from May 1956 to March 1959. At the time of writing, despite much searching, the whereabouts of the second book remain undiscovered. If anyone should come across it in the future, it would be appreciated if it could be delivered to the Dowty Group Secretary at Arle Court.

Very little original source material has been kept by the present company, which is unfortunate but not surprising. Only in recent times have we begun to take active measures to preserve and chronicle the history of our most important institutions, thereby ensuring that future generations will have a better understanding of the past. The Dowty Rotol Company Secretary, Mr Ian

Bailey-Scudamore has the original Visitors Book and a photographic album showing the building of the factory and offices. The present Works Engineer, Mr D. Biggs, kindly lent me many photographs and drawings of the buildings, dating back to the formation of the company. Most of the original wartime buildings have changed their identities many times over the years, both inside and outside, to the extent that unless one has the services of a competent guide, it is becoming difficult to remember the various offices and workshops. At the time of writing, the company have announced plans to replace many of the older buildings during the next few years, after which time we will only have plans and photographs to recall how it was. Fortunately, the Works Engineer's Department are aware that historians will want to know what the original lay-out of the buildings looked like and the great collection of plans they retain in their cabinets will prove of inestimable value.

Having set out to turn up as much from the early days as possible and with the co-operation of the present company management, I have approached the County Record Office in Gloucester in order to set up an archive for important material relating to the period, so that future researchers will be able to look back and make their own assessments. I am pleased to report that the Gloucestershire CRO were only too glad to have the opportunity to increase their existing though meagre archive of one of the major employers in the county. If anyone has any further material concerning Rotol or British Messier the County Record Office will be pleased to receive it. Hopefully, this will prevent material being thrown away after a few years, either to make room for modern papers or because nobody knows what else to do with it when retirement comes round.

Anyone who produces a book of this nature owes a great debt of gratitude to many people and in recording my heartfelt thanks to everyone who responded to my request for information I have singled out a few who contributed far more than I could reasonably expect in answer to my many queries. This does not mean that I am ungrateful to the others, but these individuals deserve a special vote of thanks.

At the head of the list is L.G. Fairhurst or LGF as most of us know him. Although there are obviously many colleagues who knew him as Len, I do not include myself among them! As a very junior draughtsman in 1953 it would have been quite unthinkable to address him in any way other than Mr Fairhurst and even though we have conducted a long correspondence over the past months, this habit has continued. Despite his present difficulties, he has given invaluable help and advice on almost every aspect of Rotol life. His memory for little details is quite exceptional and in many cases I have simply copied his recollections, with only the most minor changes. His position in the company meant that he was privy to information which affected the lives of most of the employees and he saw a larger part of the overall picture than most. All this has resulted in my being given an insight into the day-to-day working of Rotol at a high level, and enabled a much clearer view to be taken than would otherwise have been possible.

Second only to LGF in the extent of information he has supplied, is R.T. Elmes, but in his case I have been able to cross the great divide and use his forename. His letters have enabled me to include many anecdotes from his years at Rotol and some of the following chapters would be much reduced without his contributions. Charles Bracher and Vernon Christie both gave particularly valuable help and advice. Ted Edwards, Bryan Greensted MBE and Squadron Leader Denis Dickson enlightened me on aspects of the Flight Test Department that were causing some confusion and Miss Florence Jackson was helpful with details of her work during the war for the Rotol Controller, R.H. Coverley. Mr L.S. Armandias and Alan Cameron-Johnson gave considerable advice with the chapter on British Messier. Ivor Brooking, Doug Grazier, Arthur Hayward and Denis Savory produced several pages of information about their early days with the company, all of which was invaluable. Clarification on the flying aspects of certain parts of this book were either corrected or supplied by Jack Meaden, a close friend with a lifelong involvement in aviation. He was also kind enough to undertake a critical look at the work in its final draft form and made many suggestions for improvement.

Among the present-day employees at Dowty Rotol who have helped in a variety of ways I must mention Hilary Saxby, Sandra Whitehead, Angie Swain, David Jones and Dave Nixon, all of whom were instrumental in locating interesting items from the past. Patrick Hassell, a comparative newcomer to the firm, was anxious to offer his help and supplied some useful details about the Bristol Aeroplane Company. Also in this connection I thank Bob Bass for his contribution to the historical aspect of propellers. Many of my past colleagues in the offices and factory have given suggestions concerning important events or people I had overlooked and finally I offer my thanks to the management at Dowty Rotol, in particular to Mr D.G.M. Davis, the Technical Director, for encouragement throughout the months of research spent on this book.

To the following, all listed in alphabetical order, I extend my most grateful thanks for everything they have done to make this book what it is. I only hope that reading the results of these labours will give them as much pleasure as I have had, not only in researching and writing it, but also in the opportunity it gave me of meeting so many old friends and colleagues. If your name has been omitted from this list, due to an oversight, I tender my sincere apologies:

D.F. Ancell, R.J. Armstrong, S. Bartlett, R. Base, Mrs J.E. Berrell (née Bradshaw), M. Berrisford, Mrs E.M. Bloomfield (née Goulter), D.N. Birrell, D.R. Bishop, J.E. Boast, S. Booth, L. Boyd, G. Brazil, R. Brown, P. Carlile, J.A. Chilman, R. Copper, Mrs M. Cousins, E. Cowley, E.J. Cracknell, J. Dancey, C. Davey, D.L. Davis, E. Dyer, W.G. Elliott, E.G. Fairhurst, G.N. Folley, J. French, G. Gardner, J.I. Golesworthy, W. Green, N. Grinnell, A. Gunn, T. Hammond, C. Harmer, G. Hart, B.V. Holberton, F.B. Holtham, F. Humphries, Sir Robert Hunt, J.O. Hunt, Mrs E. Hunt (née Alder), W.E. Hunt, Mrs D. Hunter, W.R. Hurcombe, A.D. Ingram, Mrs L. Hill (née Blackwell), D. Jackson, C.F. James, W.E.M. James, Mrs J. Jarvis (née Higgins), L. Jones, M. Kent, G.F. Kent, F. Loveridge, W. Maclean, R. Marfell, E.P. Marshall, J.G. Martin, F.E. Morris, R.C.M. Neale, Miss V. Nelder, Mr and Mrs D. Parker, G. Pettit, G. Phillips, J.E. Pickford, R.A. Pirie, S. Price, P. Pryse, C.E. Purnell, R. Ray, S. Reed, D. Rendall, H. Reynolds, N. Robertson, Mrs J. Robertson (née Hart), Mrs E. Robjohn (née Walker), K. Robinson, G. Ronson, R. Russell, Mrs N. Ryder (née Bound), Mrs E.G. Sandford (née Williams), R. Seeley, E. Smith, J.D. Smith, L. Smith, P.R. Smith, D. Soley, C. South, D.J. Spencer, Mrs M. Stafford (née Trapp), A.E. Stockdale, J.H. Sullivan, H. Taylor, H. Twitchen, J.P.W. Turner, N. Tyrrell, Mrs C. Wain (née Wigley) J.G. Walter, B. Watts, R. Watts, W. Webb, Mrs S. Weller (née Beasley), D.E. White, D.F. Williams, H.J. Williams, R.C. Williams.

Sadly, a few of those listed above are no longer with us, having passed on in the intervening period during the months of research and subsequent publication.

The illustrations have come from many sources. Although the negatives seem to have disappeared with the passing of time, the present company have a large collection of prints, only a small number of which are reproduced here. A few personal photographs have been kindly loaned by individuals, mostly taken by Don Williams the official Rotol Photographer for many years. The rest are copies from brochures, periodicals and magazines of the period.

In an early edition of *The Airscrew* magazine I came across a passage, written by the then Works Manager, Mr Arnold Jackson, which I can do no better than quote in full, for it seems to epitomise not only the hope for all of us in the dark days of wartime Britain, but also my personal hope that the reader will enjoy the results of my research into life at Rotol.

'One day, please God, there will be an end to this war and perhaps after some long time we may find a copy of this friendly little book (*The Airscrew*) and blow the dust off it, read it again and remember events that interested us and people who were our friends.'

Early History of Propellers

The history of Rotol Ltd (formerly Rotol Airscrews and now Dowty Aerospace Gloucester, part of the Dowty Group of companies) began in 1937 when the Bristol Aeroplane Company Ltd and Rolls Royce Ltd. set up a joint organisation to cater for propeller development. The word Rotol was chosen as a shortened amalgam of the two companies. The reason for their decision in submerging normal company rivalries will be found in the proceeding chapters, but a knowledge of propeller development leading up to 1937 is useful, although not essential. (The term 'airscrew' and 'propeller' are interchangeable, although the use of the former has tended to disappear with the passage of time.)

I am indebted to Mr R.M. Bass for granting me permission to draw freely from his paper published in the Aeronautical Journal dated August/September 1983 entitled 'An Historical review of propeller developments'. Bob was a propeller design engineer at Rotol for many years and retired in 1985.

The propeller as a means of propulsion has been in use for a considerable time but for the sake of brevity we need only consider the relatively short period when it has been fitted to aircraft. In the Great War, usually known nowadays as the First World War, the aircraft propeller was carved from wood and had a fixed pitch, although it was realised that one which could be varied in pitch to cater for different engine speeds would offer a far greater performance. Attempts had already been made towards this end and during the latter half of 1917 a four bladed propeller with variable pitch, manually controlled by the pilot, was made and tested at the Royal Aircraft Factory . The admittedly limited success of this venture showed that a variable pitch propeller would show a great improvement in aircraft perfomance and in 1918 a two-bladed version was fitted to an SE5. Although there was still much development work before a satisfactory VP propeller would become a reality we may consider these humble beginnings to be the start of present day conditions in the propeller industry.

Direct manual actuation of blade pitch by the pilot was not satisfactory and a hydraulic solution was sought. In the mid-1920's Dr H.S. Hele-Shaw and Mr T.E. Beacham patented such a device in which hydraulic actuation of the blades was described. The drafting of this patent coincided with the general development of supercharged engines which required variable pitch airscrews in order to fully exploit their potential. In 1925 the Air Ministry invited tenders for an all-metal VP airscrew and in August tests were carried out at RAE Farnborough. The first constant-speed propeller, with two Leitner-Watts hollow steel blades made by Metal Propellers of Croydon and a hub made by Harper Sons and Bean of Dudley, was fitted after ground testing to a Rolls-Royce Condor III engine and subsequently flight trials were carried out.

Metal blades had been chosen because of the difficulty experienced in attaching

1920's only natural hardwoods were available for propeller manufacture and it was only with the advent of compressed resin-bonded composite woods that this problem was finally overcome.

The Air Ministry followed these successful tests with a development contract for twelve propellers, to be produced by the Gloster Aircraft Company. The development was undertaken by a man who was destined to play a vital part in the Rotol story.

Harry Lawley Milner (but always 'Pop' to his many friends) was born in 1886 and received his education, first at Manchester School of Technology and then at Manchester and London Universities, following the award of the Whitworth Scholarship. He served an apprenticeship with Sir W.G. Armstrong Whitworth Ltd before becoming engaged on aero engine design at RAE Farnborough from 1921 to 1922 when he joined Hele-Shaw and Beacham for a short time until moving to Gloster Aircraft in 1924, where he became closely involved in propeller design and with the Air Ministry contract in particular.

Two propellers were developed and flown in 1928, one with hollow steel blades fitted to a Rolls-Royce Kestrel engine and the other with solid duralumin blades to a Bristol Jupiter. Limited testing was carried out on the Kestrel propeller but that fitted to the Jupiter had over 250 hours flight testing in Gloster Grebe and Gamecock aircraft. The pitch range was in the order of about 12°.

In the next few years engine powers and aircraft speeds increased to such an extent that fixed pitch propellers became unacceptable and although there were many who remained sceptical, variable pitch was forced on an initially reluctant aircraft industry. Hollow steel blades ultimately proved unsatisfactory and the task of finding a suitable alternative was given to the two major engine manufacturers, Rolls-Royce and the Bristol Aeroplane Company, under sub-licence from the Gloster Company. These two companies set up Rotol Airscrews Ltd. to develop and manufacture propellers and H.J. Milner joined the new company and remained with it until his death in December 1956.

It would be invidious to ignore the activities regarding propeller progress in the United States where Hamilton Hydraulic Airscrew Company, following a visit by Hamilton to GAC to view the Gloster built VP propeller, applied for a patent for a two-position propeller in May 1929, four and a half years after the Hele-Shaw Beacham British patent. Progress however was much swifter in America and their propeller went into production in 1934. In the meantime, another British company, the De Havilland Aircraft Company acquired a licence from Hamilton's to manufacture either a fully variable pitch constant speeding or alternatively a two position propeller. De Havilland made such good progress with the Hamilton design that by 1939 nine out of every ten Mercury or Pegasus engines produced by BAC were fitted with these propellers. By the outbreak of war, Britain and America had four major propeller companies, the only one not mentioned so far being Curtiss-Wright, who were making a propeller with electrically operated pitch change which, at a later date, Rotol made under licence.

Meanwhile, events in Europe were having a marked effect on aviation as the growing Luftwaffe tried out its new aircraft in Spain during the Civil War of 1937. German propeller manufacturers had not been slow in adopting the growing Luftwaffe tried out its new aircraft in Spain during the Civil War of 1937. German propeller manufacturers had not been slow in adopting the advantages of variable pitch to their aircraft, indeed it was to come as an unpleasant surprise to the Western Allies in the next few years when they began to discover just how far ahead the Nazi war machine was in almost every sphere.

Before progressing to the history of Rotol in the chapters which follow, the reader may wish to refer to 'Thrusting Forward', a book written by George Rosen with Charles A. Anezis, which concerns one of Rotol's rivals. It was published by the American propeller company Hamilton Standard, in collaboration with British Aerospace Dynamics Group (Hatfield-Lostock Division), to coincide with the fiftieth anniversary of the formal association of these two companies, although in 1934 it was the De Havilland company which figured on the original letter headings. The book outlines the progress of the propeller from antiquity to the present day, and is fully illustrated with many excellent photographs (naturally with the accent on the Hamilton Standard propeller), and anyone with an interest in the subject will find it most fascinating.

The Birth of Rotol Airscrews

Although his name is rarely mentioned in this connection, one man's foresight, at a time when few others were convinced of the value of variable pitch propellers, was principally responsible for the creation of Rotol Airscrews Ltd. That man was Roy (later Sir Roy) Fedden, Chief Engineer at Bristol Aeroplane Company from 1920 to 1942 and a well respected figure in aviation. His tenacity and vision played a major part in ensuring that the British airscrew industry was in being and well established by the time the war started, without which the fighter aircraft of the RAF would have been at a considerable disadvantage during the Battle of Britain in 1940.

Great strides had been made with aircraft engine design in the 1930's, and Fedden was very conscious of the need for variable pitch airscrews, which were obviously necessary in order that further advancements could be made. He was aware that a small team at Gloster Aircraft Company were engaged with the problem but despite his frequent and persistent enquiries to Harry Folland, their chief designer, as to the current position of the Hele-Shaw Beacham airscrew, such progress as there was did not satisfy Roy Fedden.

Irked by the apparent lack of urgency being shown, he attempted to persuade his company that they should set up their own department to tackle the problem but Bristol Aeroplane Company were unable or unwilling to spare the resources. In desperation, Fedden took the almost unprecedented step of attempting to persuade the De Havilland Aircraft Company at Hatfield to take an interest in the range of airscrews produced by the Hamilton Standard Propeller Company of America, in order that Britain should not be left out in the cold altogether. His approaches were successful and shortly after, De Havilland began to market them under licence.

Fedden did not simply sit back and leave De Havilland to get on with their licence production. Working on the assumption that competition was good for industry, he next took his ideas to the Air Ministry, where he had the sympathetic ear of Air Marshal Dowding, the Commander in Chief of Fighter Command. Dowding asked Fedden to produce two airscrews to the Hele-Shaw Beacham design, for assessment by the Air Ministry, and a delighted Fedden wasted no time in convincing Gloster Aircraft to help him set up a design team at Bristol. These included Hugh Burroughs, Harry Folland and H.L. ('Pop') Milner, and within a short space of time they had produced a variable pitch airscrew for fitment on a Mercury-engined Gloster Gauntlet, which successfully completed a 100-hour test programme.

Despite the success of the Bristol/Gloster team, the BAC board remained unconvinced that there would be a market for another manufacturer, De Havilland's having made such progress with the Hamilton design, and this might have been the end of the story, were it not for Fedden's persistence. The decision he came to at this point must have caused Fedden some heart searching. He invited Arthur Sidgreaves and E.W. Hives, two members of the rival Rolls-Royce company, to lunch with him at the Royal Thames Yacht Club, to discuss the idea of setting up a unified modern propeller production programme, which would be of benefit to both companies. Before lunch was over they had reached broad agreement, and to Fedden's surprise, the Board at BAC also accepted the idea on his return.

The Ministry of Supply, conscious of the threat of war with Germany, and realising that it would be in the national interest to concentrate the talents from Rolls-Royce and Bristol Aeroplane Company into a single organisation, lent their good offices in the talks which followed. The results of these discussions was the setting up of a company which would develop and manufacture variable pitch airscrews for the new range of aircraft engines now being designed.

Rotol Airscrews Ltd came into being on May 13th 1937. Despite the rivalry of the parents, the 'marriage' seems to have worked quite well, although there were occasional moments of anger and high drama behind the scenes. The Rotol executives who dealt with the day-to-day running of the company had to exercise a certain amount of tact and diplomacy when confronted with these situations, as might be imagined.

The question of standardisation in the propeller industry also required urgent consideration. Prior to 1937, little progress had been achieved regarding the size of propeller shafts and hubs and this was obviously the starting point for any future development. The SBAC Standards Committee were requested to formulate some realistic proposals, which would be acceptable to the various companies involved in aero engine and airscrew design. The result was a set of standards to cater for a number of airscrew shaft sizes throughout the engine power range, together with standard flanges for the engine-mounted constant speed units and feathering pumps. In order to prove that the companies could achieve interchangeability of their airscrews using the new standards, a demonstration took place at the Bristol Aeroplane Company on December 13th 1937 in front of Air Ministry officials, at which both Rotol and De Havilland airscrews were successfully interchanged. This enabled airscrews from either of the two companies to be fitted to the aircraft engines being produced for the RAF and was to be of inestimable value during the Second World War. Eventually, the SBAC went on to produce standards for the mountings and drive shafts for auxiliary gearboxes, which still lay in the future.

The base chosen for the new company was in the ancient city of Gloucester, presumably because it lay between Derby and Bristol but other sites may have been considered. Rotol's first address was the Bell Hotel, Gloucester, where a room was rented for two or three weeks, while negotiations for temporary office and factory premises were carried out.

Mr R. Stammers was appointed as General Manager, a post he was to occupy for several years. (The choice of name for the new company – RO from Rolls-Royce and TOL from Bristol, is credited to Mrs Stammers.) His early training was in civil engineering and before the First World War he had been engaged on dock construction work. After service with the Corps of Royal

Engineers, first in France and then in Salonika, he was demobilised from the army with the rank of Captain and immediately joined Roy Fedden as his personal assistant. Fedden was at that time Technical Director of Cosmos Engineering, the company who were the forerunners of Bristol air-cooled radial engines. When Bristol Aeroplane Company formed its Engine Department, Stammers became their Commercial Manager in 1920. He was considered to be the perfect gentleman, indeed according to one of my informants, 'Stammers was so much of a gentleman, he even held doors open for office boys.'

The second appointment to be made by the new company was that of Company Secretary. As Stammers had come from BAC it was almost inevitable that this post would be filled from the ranks of Rolls-Royce and the choice fell to the Assistant Accountant, Mr E.O. Cameron. He was somewhat younger then Stammers and had qualified as a Chartered Accountant around 1920, joining Rolls-Royce in 1924, where he had taken a great interest in the Welfare Society. He played a leading part in political work in Derby, although we are not told to which faction he belonged, even standing as an unsuccessful candidate for a local council election. Cameron became both Secretary and Chief Accountant for Rotol and in later years, recounted the tale of how one of his first tasks was to persuade the local bank manager to part with ten pounds to enable some purchase to be made!

The share capital, divided equally between Rolls-Royce and BAC was £250,000, increased to £500,000 in 1939 and £600,000 in 1959. The choice of directors, two from each company, was fairly obvious, given the enthusiasm already shown, at least by the Rolls-Royce management. The original members of the Board of Directors of Rotol Airscrews were; Sir G. Stanley White Bt, A.H.R. Fedden MBE (from Bristol Aeroplane Company), A.F. Sidgreaves OBE, E.W. (later Lord) Hives MBE (from Rolls-Royce). These four, together with W.G. Verdon-Smith, Stammers and Cameron met for their first Board meeting on June 3rd 1937, in the temporary premises which Stammers and Cameron had leased in the docks area of Gloucester. These buildings, which had once housed a carpet factory, are still in existence at the time of writing. The offices fronted on to Llanthony Road, with the workshops running at right angles down High Orchard Street.

At this meeting, at which Sidgreaves was elected chairman, it was resolved that the directors should retire in alphabetical order and that, for the moment, no remuneration be paid to any of the directors. Other matters under discussion concerned the setting up of offices for the Drawing Office Staff and Technical Department, whose arrival was imminent. It was anticipated that the staff from Bristol would move to Rotol in the week commencing June 7th. Captain Frank Barnwell, who had been responsible for the majority of Bristol designs since the First World War was retained as Airscrew Consultant, at a fee to be agreed between himself, White and Fedden. His involvement at Rotol was not to last long as Barnwell was killed in 1938, flying in a privately built light monoplane.

The directors also received a report from Stammers on the current situation regarding negotiations for the proposed new Rotol factory at Staverton. The premises at Llanthony Road had never been intended as anything more than temporary accomodation and Fedden and Haldenby of Rolls-Royce apparently considered several sites, including one at Shrewsbury, before deciding on Staverton. Not only did it have the advantage of being adjacent to the recently opened Gloucester Municipal Airport, but it was situated about halfway along the main road between Cheltenham and Gloucester, making it easily accessible to an increased workforce. One other factor was the new housing estate being built in the immediate vicinity, in the village of Churchdown.

The Cotswold Aero Club, who had been flying from the grass field, which was now the proposed site for the new factory, were persuaded to transfer their operations to the aerodrome across the road. They were not too unhappy with the change, as the aerodrome was much better drained than the Rotol site, where the Hatherley Brook sometimes overflowed in the rainy season – and sometimes out of season too! Indeed, many of the Rotol employees will remember the occasional 'high tides', which were quite spectacular to behold and were only cured by the use of extensive and expensive mechanical ditching in the 1970's. On at least one occasion, the force of water was so great that it even lifted manhole covers, and some original drawings that had been stored in sheds near the brook floated away down to the River Severn. Subsequently, if a drawing could not be located, it was usually dismissed with the comment, 'It must have got lost in the flood.'

A wooden building used by the Cotswold Aero Club was retained as the first Rotol Club House and an unfinished house, also belonging to the Aero Club, was converted into two flats for company employees. The Fireman and Maintenance Engineer were suggested as possible tenants, at a rent of 12/6 (62p) per week.

Instructions to commence work on the new factory were issued at the first Rotol Board of Directors meeting on June 2nd 1937. Messrs Gardiner of Bristol fabricated and erected the steelwork for the offices, Robert Watson of Bolton were responsible for the steelwork in the factory, while Higgs and Hill of London, as the main contractors, prepared the site, laid the foundations and looked after the remainder of the building work. The construction of the new factory and offices seems to have gone ahead at a furious pace. We are fortunate that someone had the foresight to engage the well-known photographer, H.F. Jones of Gloucester, to make a weekly visit to the site and take a series of excellent quality pictures, both outside and inside the factory, which give us an insight into the rate of progess which was achieved. Judging from these photographs, 1937 was a good year, as the sun is often shining and the skies are filled with white clouds. The speed with which the whole building was accomplished indicates that the weather did not cause any hold ups, and by the end of July, all the steelwork had been erected and much of the brickwork was going up. A surprisingly small work force is evident in the pictures, and the only sign of mechanisation is a large crane and a cement mixer. It is an indication of the times that with the aid of a magnifying glass, a horse and cart can also be seen.

In the meantime, on May 13th 1937, new employees were beginning to turn up at Llanthony Road, most of them transferring from Rolls-Royce or BAC with the inducement of higher pay or increased responsibility or both. The honour of being the very first to arrive went to Ron James, who eventually became the Maintenance Superintendent, but a number of others, also from BAC turned up later the same morning. The little factory began manufacturing operations at the end of May, with a small number of operatives, probably only just into double figures. C.J. Luby was the Works Manager, also ex-BAC although the exact date of his appointment is not known. Reg Barrell, who had been in charge of the Experimental Airscrew Shop at BAC when the Bristol Hele-Shaw airscrew was made and tested, transferred to Rotol to take charge of the Machine Shop and in February 1938 became the Chief Inspector.

No records have survived – if they were ever written – of the precise manner in which the initial staff were chosen for the new company. Discussions took place at both Rolls-Royce and BAC regarding appointments to be filled at the new factory, and selected individuals were told of positions which were available. The chance of being in at the beginning of a new enterprise must have been an attractive proposition, which would not have been turned down lightly.

This was certainly the case for Ernie Smith, who was called into the office at Derby and told by Hives to 'go down to Gloucester and start an Inspection Department at Rotol.' On August 17th 1937 he set off with a Third Class rail ticket for an interview with C.J. Luby and following confirmation of his appointment Ernest reported at Llanthony Road on August 30th 1937, where his first job as head of the Inspection Department was to advertise for staff in the local newspapers. The result was a sore disappointment. Of the 300 applicants, only six were considered suitable. This was to be a recurring problem, and the search to find the right people for the new factory went on for many months and extended all over the country.

Ernie Smith shared the compartment on the train journey to Gloucester with two other Rolls-Royce employees, L.G. Fairhurst, who had an appointment with Stammers, and his younger brother, E.G. Fairhurst, who was to discuss development issues between BAC and Rolls-Royce. L.G. Fairhurst had started his career as a premium engineering apprentice with Leyland Motors, moving around the North of England a great deal before joining the Aero Engine Division of the Bristol Aeroplane Company. After three years in the Advanced Projects Office, working with 'Pop' Milner on the Bristol hydraulic internal cylinder airscrew and also on Curtiss-Wright electric types, in 1935 he moved once more, this time to Rolls-Royce. Here, in an unguarded moment, he criticised certain features of a variable pitch airscrew design, and was rewarded for his outspoken comments by being transferred to work on hydraulic airscrews, under Hele-Shaw Beacham patents.

Some initial confusion surrounded his transfer to Rotol Airscrews. Apparently, LGF went to Llanthony Road under the impression that he would be the Chief Hub Designer for the new company, this being the position selected for him by Hives at Derby. On his arrival he found to his dismay that the Bristol Aeroplane Company had already given this post to 'Pop' Milner. Bitterly disappointed, he immediately applied for and was given the position of Chief Designer for Rotol's great rivals, De Havilland's. He handed in his notice to Stammers at Gloucester, but before he could disappear from the scene he had to report to Hives, who no doubt told him not to be so hasty, as the situation could change with time. He accepted these terms, and the rest, as in most good stories, is history. On his return to Gloucester he took on the position of Chief Draughtsman, but went on to become the Chief Engineer and eventually Technical Director at Rotol, for almost the whole of the period spanned by this book and retired from the Dowty Group in 1964, aged 60.

Possibly a similar story can be told for some of the other major appointments, as both parent companies put forward the names of those considered most suitable. This led inevitably to a certain air of rivalry between the employees, coming from such prestigious backgrounds as the two major British aero engine companies, each of whom thought – and in some individual cases were heard to declare – that 'their way was best'. There was a saying at the time that 'Gentlemen came from Bristol's but Rolls-Royce produced engineers' and this attitude survived for quite some years, indeed it can be said to have existed at some level in the company until taken over by the Dowty Group in 1959.

The list of Llanthony Road personnel in those first few weeks, reads something like this: Cliff Smith (Blade Shop Foreman), B.S. Jelley (Plant Engineer), H. Newport (Development Engineer), W. Brierley (Chief Blade Designer), H. Milner (Chief Hub Designer), L.G. Fairhurst (Chief Draughtsman). In the lower echelons were Bill Peel (Accountant), Albert Adams (Cashier), M.H. Stephenson (Cost Accountant) these three were all from Rolls-Royce and Miss Goulter (Cameron's secretary), Jim Finch, who, after a short career in the Drawing Office, had the distinction of becoming Rotol's very first Resident Technical Officer (RTO), and Messrs Hawkins, Chambers, Bill Nottingham (later to become Production Engineer), Tingle, Ted Marshall, Jack French, Harold James, Miss Fearn, Jack Bettridge (the first Shop Steward) and E.J. (Mac) Cracknell, who was the very popular Despatch Superintendent in the years that followed. This list cannot be considered complete as no doubt other personnel were engaged throughout this initial period.

J.H.M. Parker was one who joined the new company in those early days and stayed with them until retirement. Although he held a Master of Arts degree he was a rather shy and retiring individual and initially was given the task of setting up the records in the Drawing Office. This resulted in the evolution of a card index system for each drawing in order that a proper history could be kept of the parts being made in the machine shop, and because Mr Parker took his job most conscientiously it was his custom to take the card system home with him when he left work each night. Unfortunately, on one occasion, when he was alighting from the bus in Gloucester his suitcase flew open, scattering the precious records all over the ground. From then on he left them in the office.

Jack Dancey started work in the Llanthony Road office as a Junior Clerk, (Clock No 4) at the end of May 1937. His letter of engagement confirms the staff hours of work at that period, which were 9.00 am–1.00 pm and 2.00 pm–5.30 pm Monday to Friday, giving a total of 37½ hours per week. As the junior in the office, he did the general running around, at the beck and call of everyone, even cycling to the Post Office at the end of each day to get letters and packets into the mail, in those days having to buy the postage stamps with cash from the 'office float'. Jack made the move to Staverton with the rest of the factory at the end of 1937, and was called up for army service in January 1940.

A story circulated at the time was that local residents were under the impression the factory were producing a form of woodscrew, because of the name 'Rotol Airscrews'. The term 'airscrew' was not universally recognised as a form of propulsion and it was only when they saw the large propellers coming out of the gates that this misapprehension was dispelled. On the other hand it may simply be one of the many folk legends relating to Rotol.

Although there was plenty of space for the machines at Llanthony Road, as can be seen from the photographs, the office accommodation seems to have left much to be desired. One of the two lady tracers, Joan Hart, (eventually to marry a Scotsman, Norman Robertson, who joined Rotol about a year later) recalled that she and Rita Few (née Burley) frequently complained at the inadequacy of the arrangements, as their drawing boards occupied very cramped conditions on the first floor landing.

The Transport Department consisted of two vehicles, a 20 hp Flying Standard and a 15 cwt Bedford truck, and two drivers, both of whom were still with the firm 20 years later. It seems to have always been company policy to transport their products by road whenever possible, although De Havilland are reputed to have used rail services for their propellers.

Under the direction of Reg Barrell, a number of airscrews were produced in the few months of operations from the Gloucester premises. The early blades were made from magnesium alloy forgings supplied by High Duty Alloys of Redditch. These were fitted to a Hawker Hart K5044 powered with a Bristol Pegasus X engine and with this combination the take-off and rate of climb were hailed as 'phenomenal' by the aviation press, with a top speed of over 200 mph. This was the very first variable pitch propeller to pass an Air Ministry Type Test in Britain. The SBAC Flying Display in June 1937 was also the first time that Rotol were associated with this British institution, when the Hart, now fitted with a Mercury engine driving a 12 ft 9 in diameter Rotol airscrew, created quite a stir because of its much improved performance.

The new company were also producing adjustable pitch airscrews for the Gloster Gauntlet and Bristol Bombay at this time, although any adjustment that was required had to be done by hand on the ground. The roots of the individual blades were calibrated and fitted into a split hub, the blades then being set to a datum mark on the hub by slackening off the securing bolts. A start was also made on the airscrews for the Vickers Wellesley bombers of the RAF Long Distance Flight, which will be related in the next chapter.

Fedden and Stammers attended a meeting with officials of the Air Ministry on 12th July 1937, during which agreement was reached to order 500 hydraulically operated airscrews suitable for 'Mercury Blenheim' type and 250 for 'Merlin Battle' type, although the Air Ministry officials were at pains to point out that they did not expect to pay a higher price for the Rotol airscrews than they already paid for those supplied by the Bristol Aeroplane Company. These airscrews were to be supplied with magnesium blades, but it was anticipated that a total of 100 might be required to have wooden blades. In order to be in a position to supply wooden blades, Stammers was instructed to approach the Jablo Company with a view to making the Heine blade under licence, exclusively if possible, but on whatever terms could be negotiated. (Details of the different wooden blades will be found in a later chapter.)

Finance in those early days was clearly something of a problem, with costs steadily mounting but no income yet forthcoming. At the Directors meeting in

July 1937, the company secretary reported that the bank balance stood at £6,000 and that he would shortly be required to pay out £49,377. This figure included £11,000 for the new building at Staverton and £22,500 for machine tools from the Bristol Aeroplane Company. An urgent call for help was made to the two parent companies, to remit an immediate loan of £25,000 each.

Meanwhile, the company waited anxiously for the promised Air Ministry contract. When it materialised, sometime in late autumn 1937, it was for only half the expected figure of 750 airscrews previously quoted but it was anticipated that the remaining 375 would be ordered in the next financial year. The Ministry contract was originally with the Bristol Aeroplane Company but transferred to Rotol in order to ensure that the new company was in a position to commence manufacturing on a production basis as quickly as possible, and Rolls-Royce are also believed to have shifted some work to Rotol from their Derby factory for the same reason. This was obviously sound policy, for without it a great deal of time could have been lost, but as the weeks went by it was also to be the cause of some doubts as to the ability of Rotol to stand on its own feet. At several meetings of the Directors, anxious voices were raised at the seeming lack of progress on the part of those responsible, to obtain contracts on their own volition.

Enquiries for a large quantity of airscrews for the Supermarine Spitfire had been received in the summer of 1937, which gave rise to a good deal of discussion as to how the company would be able to fit these and other airscrews into their existing programme. The directors agreed that it was essential for the future success of the company that all orders for the new types should be accepted and worked in with the general programme, even to the extent of allowing delivery dates to slide a little if this could not be avoided. Prototype airscrews were to be handled by the newly formed Experimental Department, although the first intention of building a separate Experimental Shop on the new Staverton site was dropped owing to the cost involved. A sum of £13,000 was set aside for this new department – eventually located in No 1 Machine Shop – to include the additional machine tools which were required.

Prior to the meeting on December 14th 1937, a long report was submitted to each Director regarding the wood blade position, in which it was noted that 'Rotol were being requested – on all sides – to produce hydraulically operated airscrews with wood blades.' In an endeavour to comply with these requests, the company were conducting laboratory experiments with three materials, Heine, Halila and Permali (these were all different processes for bonding pressed wood in block form). Meanwhile, the Air Ministry had given Rotol a contract to supply three blades from Heine material, which was to be machined from blocks supplied by Jablo Propellers Ltd. One other factor which had to be considered was the delicate situation regarding patents, as all these forms of pressed wood blocks were protected against any infringements. In an endeavour to get things moving, the Air Ministry assured the company that they would indemnify them against any proceedings which might be taken, although whether this indemnification was applicable to both civil and military applications was not made clear at this stage. In any event Rotol continued in their quest for the best material from which to produce wood blades.

Sufficient machinery had been installed to enable blades to be produced for ten Mercury Gauntlet adjustable pitch airscrews per week, but delays with the receipt of magnesium forgings meant that this target was not reached. The position regarding the hubs, which were being machined by Spiers of Leicester, was even more unsatisfactory. By the end of 1937, a total of 86 blades had been completed but only 6 hubs had been supplied, against Spiers promise of 40, and Stammers was directed to find an alternative supplier. One noteworthy success reported to the Directors in December was the despatch of the first complete Rotol airscrew to Farnborough for spinning tests.

As the year drew to a close the Directors approved the payment of a Christmas bonus to the weekly staff, up to a maximum of one weeks pay depending on length of service. Those who had transferred from Derby or Bristol were treated as having unbroken service, all others were paid a proportion. This bonus, which was very much a part of the pre-war scene in most firms, was eventually to disappear following prolonged negotiations in a Dowty Rotol pay-deal in the 1960's, when it was incorporated into the normal salary paid to employees – its passing mourned by almost everyone!

Settling in at Staverton

The swift progress made by the builders on the Staverton site enabled the transfer from Gloucester to take place over a few days at the end of 1937, and the factory and offices began to function in their new surroundings soon after the Christmas break. The move was accomplished quite smoothly, with little interruption to the existing production programme.

The workforce could not fail to be impressed with the size of the new premises. By comparison with the small workshops to which they had been accustomed, it was enormous. The nine-bay factory housed the Machine Shop, which extended all along one side, with a Wood Blade Shop, Metal Blade Shop, Tool Room, Packing and Despatch Department, Stores, Plating Shop, Laboratory, Boiler House and various offices and sundry small departments, all under the one roof. Building work continued throughout the following months, one of the most welcome additions being the provision of a new canteen in July 1938, and by the next year, a further five bays had been added to the machine shop.

Both the factory and the main offices (always known as the 'White Block' because of its pristine appearance at the outset) were designed by a London architect, W.G. Philips PASI of Victoria Street, Westminster, who received 4% of the contractors final account, plus out of pocket expenses, as his fee. This arrangement was to lead to some dissension at the Director's Meeting in September 1938, when considerable disquiet was expressed at the final price for the new factory, which amounted to £97,000, an increase of £10,000 on the estimated cost.

Because the original offices are considered to be typical examples of the style of architecture of the 1930's, there is a widely-held belief that they are the subject of a preservation order. This is not the case and indeed many would argue that when the third floor and the brick office block adjoining were added, the result was a disaster for the aesthetics of the original structure.

As we shall see in a later chapter, all the buildings were camouflaged during the war. When hostilities ended, the 'White Block' was restored to its former glory, but it is surprising that no attempt was made to improve the drab contrast of the brick office block, particularly as it fronts on to the main road from Gloucester to Cheltenham. (With the promise in May 1990 of some buildings being demolished over the next few years, this unimpressive block may be in the forefront of changes.)

Back in the Gloucester factory at the end of 1937, whoever was detailed to clear up the workshops in High Orchard Street, failed to fully appreciate the hazards of magnesium shavings. The propeller blades produced up to that time were machined from magnesium alloy and the resultant swarf, together with some other factory waste, was burnt at the back of the factory. The resultant blaze was so fierce that the railway line, which served the dock area, was buckled quite badly. Had Mr Cameron been consulted, this might not have occurred, as he was well aware of the combustible nature of the material, having collected some of the magnesium shavings to add to the family bonfire during the Guy Fawkes celebrations of November 5th. Rotol eventually stopped making propeller blades from magnesium but continued its use for the accessory gearbox casings, and the operator had to take great care during the machining process, as a fire could easily ensue.

Twenty years after the above incident at Llanthony Road, the circumstances surrounding another fire with slightly graver consequences, was related to me by the man responsible. It happened during a nightshift in the 1950's, when the operator reported for work, only to find his machine completely covered with magnesium swarf, and finished gearbox main casings lying all around. As it was the responsibility of the day operator to clear the machine at the end of the shift, he asked the charge-hand to arrange for this to be done, before starting on the new shift.

Nothing happened and time went by, until the charge-hand came back and told him to get started anyway, and not hang about any longer. Although safety rules regarding magnesium laid down that work must not commence unless it was safe to do so and despite his forceful argument, he had no option but to comply with the order from his superior – but he had taken the wise precaution of getting a witness to the discussion. Sure enough, within a few minutes the swarf had 'flashed' and a considerable blaze ensued. Gloucester Fire Brigade was first on the scene, followed by the Rotol Brigade shortly after, and although the fire was quickly brought under control, the damage was considerable. Fortunately, no one was hurt, but production was badly disrupted for some time and there was an inquiry, at which the operator was exonerated.

From the small number of original employees at Llanthony Road, the company slowly began to increase its labour force. It was not an easy task and the reason why the build-up took some time was not hard to find. Apart from the difficulty of obtaining suitable labour from the locality, neither Rolls-Royce nor BAC were in favour of filling the Rotol factory with their own highly skilled machinists, as both companies had already supplied most of the management and executive staff. As Roy Fedden pointed out to the other directors, the company must expect to pay as much as perhaps £5 a week for first-class machinists, but it was hoped that the lure of a new factory in attractive surroundings might be sufficient inducement to bring forth the men they wanted.

Advertising space was taken not only in various national newspapers, notably the *News Chronicle*, but also local newspapers in towns such as Wolverhampton, Coventry and Birmingham. Special mention was made of the new housing estates being built only a mile away from the factory, near Sandycroft Farm, as one of the factors which might attract engineers and machine operators to the area.

It was sheer chance that brought some of the new employees to Rotol. Bill Elliott was a miner in Newcastle, and in August 1938 he came to spend a weeks holiday in the Gloucestershire countryside, with his aunt and uncle. He liked the area, so his uncle, who worked at Rotol, arranged an interview for young Bill, then just eighteen years of age. Six weeks later, he started working for Harold James in the Blade Shop, hammering the lead wool into the base of the blades for balancing. He couldn't make head nor tail of the local dialect and they had an even worse time with his. Army service with REME took him from Rotol in 1942, where he returned in the post-war years, but moved to Dowty in 1952. Like one or two others who also went to Arle Court soon after the war ended, he found himself back with familiar faces in the Experimental Shop, when the merger between Rotol and Dowty took place in 1959.

Another of the newcomers was Edgar Cowley, who came from Wolverhampton and in addition to Edgar senior there were three sons and two daughters, all of whom eventually ended up working at Rotol, clocking up a total of over 100 years service between them by the end of the 1950's. Edgar junior, or Eddie as he was more commonly known, was apprentice No 10, joining the company in 1938, only a few days after his father. All the family were noted for being keen sportsmen. Eddie was a fine footballer, playing for many years with local clubs. His brother Don began his apprenticeship in 1939, and also played football but will probably be best remembered as a boxer. Larry, the youngest brother, became an apprentice in 1940, and was another footballer and boxer of some note. Gwen, the eldest daughter, was selected to present a purse of money which had been collected throughout the factory, to HM Queen Mary on her visit in 1943. Sheila, the youngest, started work in the Experimental DO and later transferred to the Technical Publications Office.

Mrs Sylvia Weller (née Beasley), worked for Cavendish House before the war. She saw the advertisement in the local paper and decided to apply for the post of comptometer operator, in the Cost Office under Mr Stephenson. When she started work, on 17th October 1938, she received the magnificent sum of 35/- (£1.75) which compared very favourably with her previous salary of 12/6 (62p) at Cavendish House.

Drawing Office staff proved more difficult to recruit and was often done on a word-of-mouth basis. Two section leaders in the Main DO were Wes Tattersall and Charlie Palmer, both from the Vauxhall car company at Luton, who had joined Rotol in 1938. L.G. Fairhurst was always on the look-out for further talent and he asked them for anyone else they could recommend. Thus it was that A.E. Stockdale was surprised to be called into the Chief Draughtsman's office at Vauxhall, to answer an outside telephone call. The voice on the other end was that of LGF who asked Eric to join the other two ex-Vauxhall employees, on an increased salary. He accepted the offer, which was a few shillings more than his current salary of £5 a week and travelled to Gloucester, taking up lodgings at the same 'digs' as Tattersall and Palmer. He moved into the position vacated by Jim Finch when Jim became the Resident Technical Officer, and was eventually promoted to the position of Chief Draughtsman in the Main DO at the end of the war, taking over from Frank Blair who had been forced to step aside because of illness. Eric Stockdale continued with the company until his retirement in 1971.

It was only a few months later when another 'evacuee' from Luton turned up at Rotol to join the growing band of ex-Vauxhall workers. D.R. (Don) Bishop was used to making frequent moves, his philosophy being to change jobs in order to obtain better pay. This meant that he had left Vauxhall to take up employment with De Havilland at Stag Lane, where he worked on propellers, but an advert for draughtsmen in the new factory in Gloucestershire caught his eye and he pulled up his roots once again. When he came to Rotol for his interview he asked for £6.15.0 (£6.75) per week and was rather surprised to get it without question. Don started in the Blade Section, under Frank Bonsall, and like so many others, fifty years later could still recall his Rotol clock number – it was 82.

Harold Williams had been with the Worcestershire Regiment for over thirty years when he joined Rotol in 1938 and became the DO Print Room Supervisor. Among his duties was an instruction to ensure that no unauthorised persons were allowed entry. On his first day he had occasion to use his army training to good account – he had been a Company Sergeant Major – when he evicted a tallish gentleman from the premises. A few minutes later he had a telephone call to report at once to his superior, Eric Stockdale, where he was told that certain people were to be admitted to the Print Room without let or hindrance, among them the man he had just removed, the Chief Engineer L.G. Fairhurst.

In the early part of 1938, the Directors considered the question of joining the Employers Federation (sometimes referred to as 'the bosses trade union') but agreed that the time was not yet opportune, mainly because the company was still feeling its way, coupled with the expenditure entailed. As a consequence the Directors received a letter from the Air Ministry, pointing out that the failure of the company to join the Federation was causing a certain amount of embarrassment, as two other local firms, R.A. Lister at Dursley and Williams and James of Gloucester, had complained of the new company 'poaching' their labour. The Directors instructed the Secretary to reply that the company was still in its infancy and the time was not yet right.

There had been some good results from the efforts to attract workers – 'poached' or otherwise. At their meeting of April 6th 1938, the Directors noted that productive labour employed to date was 70 with a further 40 required and agreed there was considerable improvement in the number of employees in the Works since they last met. The figure of 99 non-productive labour was in fact 29 higher than would eventually be required, due to a large Toolroom Staff. All these figures applied to manual workers and machine operators only, the number of technical and clerical staff employed was not recorded.

Production in the new factory started quietly, as might be expected. Equipment and machines were still arriving and many of the new employees had to receive some form of training, all of which took some time. During the interviews made by the author in the research for this book, most of the trainee machinists of the early days spoke very highly of the help they received from the relatively small number of 'time-served men' who showed them how to operate the complex machines and the care needed to achieve the high degree of accuracy required on almost every part. Following current engine and aviation practice, all the components were required to be interchangeable and capable of being fitted as spares and this meant they were machined to very fine tolerances. A standard note on the drawing called for any dimension not bearing its own tolerance must be within plus or minus 'ten thou' (a thou being a thousandth part of one inch) but parts which had to fit inside bores were often machined to as little as half a thou. Inevitably, this led to a low rate of output and in the summer of 1938 the

Directors instructed Stammers to prepare a scheme showing how output might be improved. Hives gave it as his opinion that satisfactory results would not be reached until the workforce were paid by some form of bonus. C.J. Luby, the Works Manager, explained the difficulty he was experiencing in obtaining suitable labour, but hoped to have a bonus system operating by the middle of June.

Both Fedden and Hives had been responsible for most of the Air Ministry orders received to date but at a meeting a few weeks later they indicated that their direct involvement would shortly have to end. The Directors pointed out to the Rotol General Manager, the Chief Engineer and the Company Secretary, that there was a considerable amount of business waiting for a fully-feathering airscrew of the type which Rotol now possessed, thanks to some considerable help from Rolls-Royce. It was agreed that Fedden and Hives should assist Rotol in completing the current negotiations with the Air Ministry regarding the additional orders for airscrews, but subsequently, as Fedden put it, 'Rotol would have to fend for itself.'

In the summer of 1938 the fortunes of the new company began to improve, coinciding with the build-up of the Royal Air Force, which was taking place with some urgency. The variable pitch airscrew, brushed aside by many in the aviation and aero engine industry as being too complicated and heavy for the resultant gain in performance, now began to find favour. The RAF expansion programme had been instigated in the mid-1930's, when the British government had realised the potential threat from Hitler's new Luftwaffe. In February 1936, the government target for front-line aircraft, to be achieved by March 1939, was a total of 1736 in 124 squadrons, and in addition the light bombers were to be replaced by medium bombers. RAF reserves were also to be increased, all of which meant huge demands on the aircraft industry. In April 1936, a scheme had been announced to build new factories for the manufacture and assembly of aircraft and engines at government expense. This resulted in the setting up of 'shadow' factories throughout the country, many of which were run by the motor car industry.

The Munich Agreement, signed by the Prime Minister in October 1938, was destined not to give 'peace in our time' as Neville Chamberlain promised, but it did produce a further twelve months grace, when the factories could turn out the new equipment for RAF Fighter Command, replacing the obsolescent biplanes with over 500 of the Spitfire and Hurricane monoplanes. It has been argued that Germany also used that time to make preparations for the war which everyone seems to have known was coming, but without that years respite, and despite the acknowledged shame of the Munich Agreement which ceded vast areas of Europe to Hitler, there is at least a distinct possibility that the outcome of the war might have been a defeat for Great Britain in 1940.

Air Raid Precautions (ARP) were introduced by the Chamberlain Government in 1938, when the threat of war first began to be taken seriously, and instructions were issued to all British companies as to the procedure to be adopted to render their factories safe from enemy air attacks. At Rotol, the cost of the scheme was estimated at £12,230 and the government agreed that this could be met by charging the cost against sales, in accordance with the terms agreed with the

Society of British Aircraft Constructors. Among the many government requirements was the provision of a secure room away from the factory to house a separate set of all drawings produced by the company, but the location used by Rotol for this purpose is not known.

Most of the aviation industry had been making fixed pitch airscrew blades from magnesium for at least ten years and it was natural to continue with its use for the variable pitch and constant-speed types, which were slowly beginning to appear. Earlier tests however, had proved that magnesium was inferior to Duralumin, due mainly to the detrimental stress effects which occurred when a magnesium blade was damaged. Attempts were made to overcome this defect by covering the blade with a suitable material but they were not successful and the Rotol technical department turned their attention to the use of wood, the traditional material for most of the early fixed pitch propellers. Extensive research had been conducted throughout the world into methods of bonding layers of veneers, coated or impregnated with a synthetic resin, subsequently dried, stacked in hydraulic presses and finally subjected to heat and pressure. This process resulted in a material that was ideal for many industrial purposes, one of them being the manufacture of propeller blades for high performance aircraft, in the development of which Germany was known to be well advanced. Rotol used blocks supplied by two German companies for early tests and Roy Fedden and Bert Newport visited both Heine and Schwartz in the late summer of 1938, in order to glean as much information as they could.

Rotol were to eventually use three different variants of this new high density wood, with names that have come down over the years which will be instantly recognisable by most wartime ex-employees: Jablo, Jicwood and finally Hydulignum. At the outset, only the first two were in use, Hydulignum, from the Hordern-Richmond factory, came a little later, of which more anon. The problem which gave most cause for concern, that of securing the wooden blades to the airscrew hub, was finally solved when the hollow steel blade root adaptor was developed and this was the breakthrough which was needed to ensure a sound future for the new airscrews. No matter which type of wood blade construction was used, they all had the steel root adaptor.

Jablo blades were built under licence from the Heine Company of Germany, on a purely experimental basis, prior to the war. (One of the first German aircraft shot down over Britain used Jablo blades and was sent to the Rotol factory at Manchester for strip and examination in the early part of 1940.) Produced from laminated boards using 0.6 mm Canadian birch veneers, they were glued together under a combination of heat and intense pressure, giving a finished board about half its original thickness. A series of five to nine laminated compressed wood boards, glued together with casein glue, formed the block, from which the finished blade was machined. At the root end, extra laminations of graduated length were introduced, giving a blade root strength of twice that of natural wood.

Jicwood, the second of the 'improved' woods to be used, was made by the Airscrew Company at Weybridge, who were producing the German Schwartz blades under licence. Laminated, compressed-birch boards were used for the highly-stressed root portion, with the outer blade sections from spruce boards of natural density. These spruce boards were first scarfed and glued to short lengths of Jicwood material and the resultant composite boards glued together to form a block.

Wooden blades were considered to have several advantages; firstly they would be lighter than magnesium or duralumin and secondly the repair of damaged blades, whether due to enemy action or accidents – of which there were a good many during wartime – was relatively simple. In addition, it was possible to produce special types of blades without the need for expensive forming dies. One important feature, which only became apparent as the war went on and the number of training accidents increased, was the ability of the wooden blades to shatter without causing secondary damage to the engine and its mounting, or to the airscrew hub itself.

But in 1938 production of wood blades in quantity still lay in the future and the company continued to rely heavily on airscrews using magnesium blades. By September 1938, Air Ministry orders for Rotol aiscrews were as follows: 275 Mercury Blenheim, 428 Merlin Whitley, 114 Taurus Albacore, 263 Merlin Fairey (a Fleet Air Arm fighter, later to be named Fulmar). This gave a grand total of 1,080, all with magnesium blades. There was also a strong possibility of 550 airscrews for the Phillips and Powys Kestrel-engined trainer, (eventually to be known as the Miles Master) and a further 1,100 for the Merlin Spitfire, depending on satisfactory type testing.

Stammers reported to the Directors in October 1938, in response to their earlier request for increased production. He pointed out that the original programme had been for 10 airscrews per week, produced by a single shift. The new estimate increased this figure to 18, of which 12 would be fitted with magnesium blades and 6 with wood blades supplied by the Airscrew Company of Weybridge. Additionally, a further 12 airscrews would be manufactured by an unnamed sub-contractor, with the assembly to take place at Rotol. This gave a total of 30 airscrews per week which it was hoped to achieve by May 1939. Any increase in production beyond this figure would necessitate a considerable amount of further capital expenditure, both in factory space and machinery, which the Directors agreed was out of the question. For the time being Rotol would have to make do with what it already possessed.

A significant meeting took place at the Air Ministry in London on 4th November 1938, attended by Hives, Fedden and Stammers, at which a request for 25% of the AM orders for airscrews was made and agreed. This amounted to some 4,000 airscrews for the RAF and did not include those required for the Spitfire, which was currently being tested with a Rotol constant-speed airscrew. Two other points had been made at the meeting, one in connection with a de-icing device for airscrews and the other relating to contra-rotating airscrews, although it was recognised that both these items were subject to many months, perhaps years, of development and flight testing before they would attain production status.

Reference has already been made to the three-bladed airscrews for the Vickers Wellesley which had been started at the temporary workshops in Gloucester. Five of these specially converted aircraft belonging to the RAF Long Range Development Flight were fitted with the Rotol airscrew and in November 1938 two of the machines established a new world record for the greatest distance flown in a straight line, when they flew from Egypt to Darwin, a total of 7,158 miles in nearly 48 hours, giving a tremendous boost to the new company.

Rotol entered the year 1939 still relying heavily on support from the two parent companies and their staff. In the opinion of the Directors, this state of affairs had already lasted for too long, and they gave much thought as to how to stop the seemingly aimless drift and galvanise the company into some independent action. The answer was to find someone with a sufficiently dynamic personality who could take control and give the company a new sense of purpose and in order to achieve this aim, the Directors chose a Rolls-Royce employee of some standing.

Mr R.H. Coverley, a middle-aged, solidly built individual, had been Experimental Department Manager with Rolls-Royce for a number of years, and more recently Production Manager in their Car Division. The first reference to this man, who did more to change the way of life at Rotol than any other individual, and in the process gave many of its employees their first taste of what a real manager could be like, came at the Board of Directors' Meeting held on 20th April 1939.

A general discussion took place before the meeting concerning the activities of the company and the entirely new picture of its productive capacity. Rotol had originally been established at the outset to produce a mere 10 airscrews per week. With the revised urgency of the RAF Expansion Scheme it was now required to increase this to 100, with a possibility of a further increase to 150 per week in the near future. In view of this change of circumstance, the Board decided to appoint R.H. Coverley as Rotol Controller – a title he apparently chose for himself – 'as a temporary measure'. In the beginning, it seemed that Coverley would only be needed to get things moving, 'or until such time as the Rotol Design and Production organisations were in a position to carry on, without further aid from the parent companies. This period was indefinite and subject to review by the Directors.' In practice, Coverley was at the head of the company throughout the war, and might have continued for a great many more years if different circumstances had prevailed. He was always known by his initials and in this connection he could be likened to LGF but as we will discover, there was little love lost between them.

Temporary or not, RHC wasted no time in getting a firm grasp of the situation. At the next meeting on 24th May he submitted a report to the directors in which he pointed out that he had already, on instructions from Hives, contracted for an output of 7,000 airscrews, through the Rolls-Royce Subcontract Department. These would be made mainly by firms in the Manchester and Glasgow areas. The report also recommended that a further expenditure of £100,000 be made for the purpose of building a separate experimental and research department at Rotol, and this was agreed.

Coverley had a son, Robert, who also came to work at Rotol and became Works Manager in the Repair Department in Gloucester, but very little has been related to the author about young Bob, except that he is believed to have married a local girl and eventually moved to the Home Counties.

Because RHC played such a vitally important part in the Rotol story during the war years we must consider him in some detail if we are to have a complete picture of his personality. His arrival undoubtedly caused a certain amount of irritation among some of the higher management. In his position as Controller, everyone else was subordinate to him and his appointment inevitably relegated some executives, notably Stammers the General Manager, to a minor role. In addition, Coverley was first and foremost a Rolls-Royce man, with all that that implied. The men from Derby were sometimes accused of considering themselves to be a race apart, and their unswerving and often outspoken loyalty to their old company was often the cause of some resentment, and although BAC men held a similar attitude, it was to a somewhat lesser degree.

The arrival of RHC was the signal for Rotol to move into action and he galvanised the whole of the work-force by his presence and his actions. Like him or not, RHC was determined to breathe a new spirit into the company and Rotol suddenly became alive in a way that no one who was there before would have credited. Most people who knew him, agreed that he was, by any standards, a difficult man, and tended to occasionally ride rough-shod over those beneath him. When he spoke, usually brushing aside any suggestion or view which did not coincide with his own, it was as well to take note, as his word was law. Ernie Smith had known RHC at Derby and was well aware of his attitude. 'If he said "I don't like that very much" – whatever it might be – you had better be certain that it wasn't there the next time he came by.' Although he was respected as a man who could get things done, he was often feared, and this set him apart from almost everyone, both high and low. He may have had friends, but they were not among the people who worked at Rotol, with perhaps one exception, whom we shall meet shortly. If this sounds too harsh a summing up it should also be noted that so far as Rotol were concerned at that particular time, he got things moving, which was exactly what was needed in the new factory. Coverley, despite his flaws, was that fortunate combination of the right man, in the right place and at the right time.

Stan Bannister was his assistant and Miss Florence Jackson his personal secretary. In an interview with Miss Jackson, now a successful author and journalist, she described her boss as 'a mild mannnered little man – on the surface – but something of a human dynamo beneath', although others I have spoken to used much more forthright terms. She accompanied him all over the country and took dictation in every conceivable place except in aeroplanes, where it was usually too noisy. Coverley must have counted himself extremely fortunate to find someone who was not only so adaptable, but also tolerant of his occasionally boorish behaviour.

Before progressing further with the account of Rotol as the war clouds gathered on the horizon, we should look for a moment at an important part of the company that was to figure largely in future years. The Directors first recognised the need for instituting an apprenticeship scheme in February 1938, but it was shelved until later in the year, when Fedden raised the question again in more forceful terms. The main principles were finally agreed in February 1939, with a grant of £1,500 for the establishment and equipment of the school. At this time the Board were particularly anxious to make certain that all the boys trained under the scheme could be absorbed into the factory itself, which may have come about because of concern over the question of a balance between workers and apprentices. This concern appears to have overlooked the almost universal policy in the pre-war years, when youngsters would leave the firm at the conclusion of their apprenticeship in order to gain experience with other companies.

S.O. (Sam) Peebles, who will be remembered by almost everyone who ever worked at the company, was appointed Chief Apprentice Instructor. He hailed from Scotland, where he had served his apprenticeship with the Admiralty Dockyard at Rosyth, and in 1939 came to Rotol to take charge of the Apprenticeship Scheme soon after its inception. At the beginning, the Apprentice School was small, with a few machines and accommodation for about twelve boys, but more adequate facilities were soon made available in the old Transport Garage. This in turn was superseded in 1942 by a further move to larger premises consisting of a machine shop and tool and inspection rooms, and in a little over three years more than 100 youngsters had started their apprenticeships. Sam was to continue with the training of apprentices until his eventual retirement and was held in great affection by most of the several hundred souls who were delivered into his care, but more than one commented that he had a fearsome bite on occasions. Fortunately for the boys, he always had a kindly secretary who could be relied upon to give a sympathetic ear to any young miscreant. In the 1950's it was Mrs Cousins who provided the motherly instinct and smoothed the path when needed, and we will meet her later on.

Among the early arrivals in the first year of operation of the Apprentice School was R.J. (Roger) Armstrong. He started at Rotol in November 1939 on a wage of 15/- (75p) per week, and prior to taking up an apprenticeship was the office boy in the Project Design Office, situated in the 'White Block', under 'Pop' Milner, plotting graphs and – as the junior – fetching drawings etc. From their position in the Project Office, with its glass partitions, it was possible to see the comings and goings into Coverley's office and to hear the occasional outbursts. He particularly noted the way that Coverley ruled his managers with an iron hand and how everyone was exceptionally careful to ensure that everything was in 'apple-pie' order whenever RHC was around. The factory had a highly developed 'bush-telegraph' system to give warning of the Controller's movements, which gives some idea of the awe in which he was held.

After a few weeks of this highly entertaining pastime, Roger Armstrong reluctantly bade farewell to the Project Design Office and began his life as an apprentice where his subsequent academic achievements were certainly praiseworthy. While in the third year of his apprenticeship, Roger won the Royal Aeronautical Society Bronze Medal, the premier award for the highest marks attained in the whole of England, for the London City and Guilds Examination on Aero Enginering Practice. He also gained three 'firsts' and one 'second' class pass in the UEI Exams. After completing his apprenticeship, Roger went to the Auxiliary Design Office at Imperial House, where he worked under the leadership of Tommy Andrews, in the Technical Office – of which more will be related in future chapters. Roger Armstrong, who went on to become the Chief Engineer of Dowty Rotol by the time he retired in 1983, was one of a number of Rotol apprentices, too numerous to name individually, who achieved high office in later years.

Included in the company's manufacturing programme in 1939 were two forms of constant speed airscrew, fully-feathering and non-feathering, with either hydraulic or electric operation. The electric version was built under licence from Curtiss-Wright in America, in fairly small quantities at this stage, and will be considered later, but production of the hydraulic variety was well advanced. Before Rotol came on the scene, both parent companies had been busy with their own versions of a Hele-Shaw airscrew, although on slightly different lines.

The Bristol design had the cylinder mounted inside the airscrew hub and although this made for a neat and tidy solution, it restricted the blade pitch range of the airscrew to 20°. This was the 'internal cylinder' type, and the first to be put into production. In the Rolls-Royce proposals, the operating cylinder was located in front of and outside the main hub, with the pitch range increased to 35°, and known as the 'external cylinder' type. Regardless of whether it was 'internal' or 'external', the cylinder was the moving part, actuated by pressure oil supplied from the engine and controlled by the governor unit. The blades were connected to the cylinder by operating pins or links, with a fixed piston carried on a sleeve member extending through the cylinder. This design philosphy continued for some years but eventually was changed to a fixed cylinder, with a moving piston to actuate the blades.

It was realised at an early stage, that the ability to feather the blades of multi-engined aircraft in the event of engine failure, and so eliminate the heavy windmilling drag of the airscrew, would be an essential feature, and the Rolls-Royce basic design was obviously the better choice. Using the extra space afforded by the external cyinder, the pitch range could be extended to 75°, enabling the blade to be turned edge on to the direction of flight. This not only reduced drag considerably, but stopped the airscrew from windmilling and 'churning up' the engine, so possibly creating a further hazard. In addition, use of the external cylinder type enabled a common hub to be used for both single and multi-engined aircraft. The removal of a stop in the hub was all that was necessary to change the airscrew range from feathering to non-feathering.

All the early Rotol designs were internal cylinder types, with 20° pitch range, initially fitted to the Bristol Blenheim with two Mercury engines and the Miles Master trainer with a single Rolls Royce Kestrel. Later developments, still with the same restricted pitch range but with a larger shaft size were used on the Whitley bomber and the Fulmar fighter, both fitted with Merlin engines. As already recorded in the previous chapter, blades were made from magnesium alloy at this stage but because of its unsatisfactory stress properties, moves were afoot to replace this material. Development testing of magnesium and both Schwarz and Heine wooden blades went on throughout most of 1938 and 1939 with no firm decision emerging. The reason for this was possibly twofold; although it was generally agreed, both by Rotol and the Air Ministry, that a change of material was desirable, a sudden changeover might have upset deliveries to aero engine manufacturers and in addition the alternative choice of wood was considered to be something of a leap in the dark at this stage.

Duralumin would appear to be the first choice as a propeller blade material and blades from this material had already been made and continued to be made on an experimental basis at Rotol. Because De Havilland, the only other major British manufacturer of airscrews, were limited to the use of Duralumin to manufacture blades, to suit their range of American-licenced hubs, it was decreed that they would receive the majority of stocks during the war, which left Rotol with few alternatives. The American aviation industry had developed hollow steel propeller blades for many of their aircraft before the Second World War but Rotol considered them generally too heavy and although some development of hollow steel blades went on at Rotol for a while it was eventually stopped in 1941, for several reasons, one of them being the difficulty of repair. At the Director's Meeting held in October 1939, Fedden reported that half of the blades for the MAP airscrew programme would be made from Duralumin, and the remainder from either wood or magnesium. His statement presumably refers to the complete MAP programme, which would have included those being made by De Havilland.

Sir Roy Fedden, the man who was largely responsible for Rotol's creation.

H.L. 'Pop' Milner, credited with much of the early work on variable-pitch propellers.

Hele-Shaw propeller in the Science Museum.

The first temporary premises at Llanthony Road, Gloucester in 1937.

Office staff in 1937; *Left to right*: Miss Fearn (telephonist), Adams (cashier), Peel (accountant), Dancey (junior clerk), Miss Goulter (secretary), Cameron (company secretary), Stephenson (cost accountant).

Llanthony Road workshops 1937. *Left to right*: Harold James, Dick Robinson., Reg Barrell, unknown, Stan Booth, unknown, Ted Marshall.

Jack French, one of the first employees, leans on a bent blade, demonstrating the popular slogan – 'Guinness for strength'.

Steelwork of the new Staverton factory, looking from Hatherley Lane, July 27th 1937.

Rotol factory and offices taking shape two months later.

The finished office block in March 1938. The 'White Block' was originally designed to receive another storey, which it did in 1951.

Reception hall with commissionaire's chair and table below the stairs.

Drawing Office on the first floor of the 'White Block' in 1938. The two tracers are Rita Few (née Burley) nearest camera and Joan Robertson (née Hart). Tom Godden is standing at drawing board, right.

Accounts Department in 1940. Seated left are Miss Johnson, Eric Slatter, S. Bartlett. Bill Biggs (fair hair) is seated right. Note the mandatory blackout curtains on the right.

No 1 Machine Shop in 1939. Note the hubs for three-blade airscrews in the foreground.

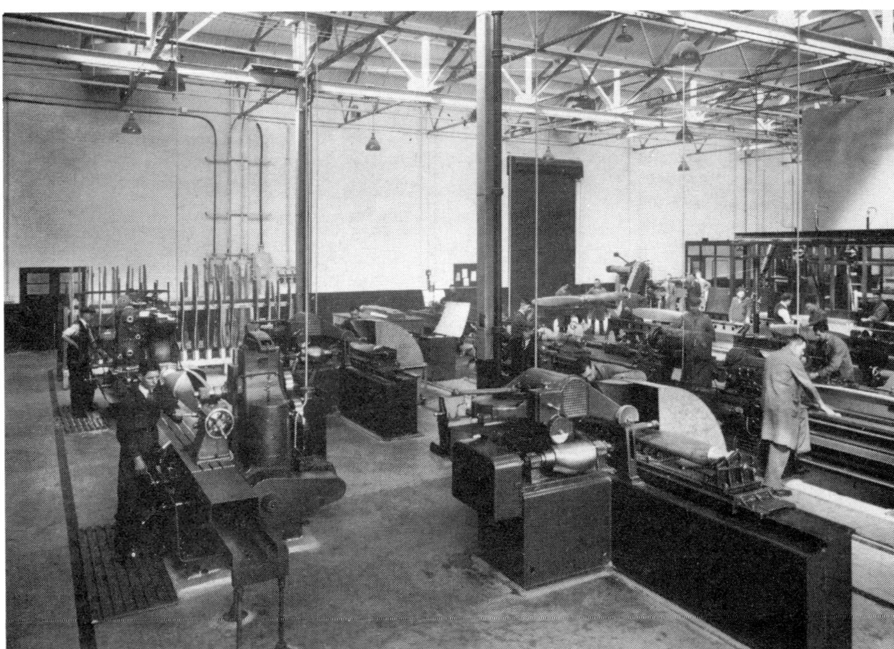

A posed shot of the Metal Blade Shop in 1939 – looking suspiciously tidy. The Berliner shaping machine for copy-milling the magnesium blades is at right foreground.

Hawker Hart K3020 powered with a Bristol Mercury, during testing of a Rotol constant-speed variable pitch airscrew in 1937.

Pilots and Navigators of the RAF Long Distance Flight with their Vickers Wellesleys. These Rotol constant-speed airscrews were part of the first small-scale production at Llanthony Road in 1937. *Via Jack Meaden.*

The Early War Period

Following the German invasion of Poland the British Government declared war on September 3rd 1939 and in common with most other other important industrial concerns, Rotol began to set up dispersal factories and offices in several parts of the country. They were generally situated in vacant factory premises, and in some cases, perhaps most, they required extensive alteration by the Works Engineers Department, for whatever changes had to be made in order to cater for the new products. As the war progressed, further factories were taken over, one of the nearest being at Worcester, where parts for Constant Speed Units were made throughout the war. Precision tools and finished parts were produced in a factory at Frome in Somerset, evacuated from Camden Town to avoid the London Blitz. Two more factories, one at Osset in Yorkshire and another near Trafford Park in Manchester, dealt solely with the production of wooden blades. The furthest of all, concentrating on electric propellers, was a factory at Kilmarnock in Scotland, and something of their activities will be related in a future chapter.

Other factories began producing Rotol components, among them the 'shadow' factories for the aircraft industry, and although there was much discussion on the subject of creating a 'shadow' factory for Rotol itself, this was never achieved. The cost of such a venture was put at £1,500,000, but the Directors chose instead to obtain additional plant for the companies already engaged on sub-contract work, at an estimated cost of £750,000. Given this increase, Coverley anticipated a further 550 airscrews a month would be made available. The figure was based on the sub-contract firms working a single shift of 47 hours per week, with the Rotol Main Works supplying 450 airscrews per month. Within a short time the machine shops began to be filled with lines of milling machines, capstan lathes, thread grinding machines, and a great many others to produce the dozens of parts needed for the finished airscrew.

Security was of such vital importance during the war, that the outlying Rotol factories and locally dispersed departments were referred to in official paperwork by their initials – thus Osset became O/F, Worcester W/F, etc – possibly in an attempt to disguise their true identity from enemy agents, although it seems doubtful if this would have fooled the German Abwehr for very long! Many of the Rotol departments were dispersed to premises in the nearby towns, but this was a gradual process, continuing throughout the war and sometimes coincided with departmental moves designed to provide extra space at Main Works, which was always at a premium.

In Gloucester, the old Crypt School was used as a store for finished blades, while upstairs became the Technical Publications Department. The Photographic Department went to the new Technical School and a large house in Brunswick

Road, 'The Lawn', was used for office accommodation. Some of the Drawing Office sections, initially housed on the upper floor of the White Block at Main Works, were moved to the Gloucester Technical College. Among them was the Electrics Section, where Charlie Palmer was section leader until Harry Twitchen took over, following some disagreement between Charlie and L.G. Fairhurst. When Harry moved to the Marine Screw Section, Ted Morris became Section Leader.

In Cheltenham too, premises had to be leased to accommodate further Rotol departments. The Press and Publicity Offices moved into 'Glensanda', a private house in the Montpellier district (now in use by the Cheltenham Borough Council as their Housing Department) while 'Imperial House', near the Gordon Lamp roundabout, formerly the offices of E.J. Burrow the printers, became the Auxiliaries Design office. (Confusingly, the house next door to 'Glensanda' is also called 'Imperial House'.) Steel's Garage in the High Street was converted into the Spinner Factory and a private house on the London Road, 'Keynsham Bank', was taken over by the Accounts and Cost Departments, becoming Keynsham Bank Office, shortened to KBO by the employees.

Almost certainly, some premises have been omitted from this list but it shows the lengths to which Rotol went in order to spread themselves and minimise the chance of the Luftwaffe dealing a knock-out blow. It also resulted in a big problem for the Rotol Transport Department, as everything had to come back to Main Works for assembly and despatch. In order to maintain communications and convey personnel from place to place now that Rotol departments were spread all over Cheltenham and Gloucester, a regular shuttle service had to be set up between both these towns. The service involved six or seven trips a day, using either Morris or Ford Utility Vans, mainly driven by women and operated to a scheduled time table, circulated throughout the various departments. (See appendix for details).

In October 1939, Rotol were asked by the Air Ministry to start an Instructional School for service personnel, to enable them to become familiar with the functioning of the new constant speed airscrews. Originally the courses consisted of a series of lectures, followed by periods spent in the assembly shops at Main Works, but the increasing numbers of servicemen from all nationalities forced a more comprehensive course and separate premises. The school moved first of all into the Spinner Factory in Cheltenham, but it was soon realised there was insufficient space to accommodate the numbers required, so another move became necessary. This time they took up residence in the City Museum in Gloucester, where the glass cases and free-standing exhibits on the floor had been removed but those on the walls and in the gallery remained. Many of those who passed through the school remember the stuffed head of a water buffalo hanging on the wall, regarding their first efforts with the pre-loading spanner! Instruction covered operation, maintenance and repair of airscrews, including wood blades damaged by enemy action or accidents. Dozens of RAF and Fleet Air Arm fitters and carpenters soon began to learn the proper techniques for plugging bullet holes and 'scarfing' new pieces on to the blades.

The move to Gloucester Museum was again only a temporary measure, and in April 1941 the school went to rooms in the old Technical College, where the intake of servicemen was increased to 25 per week. In addition to receiving instruction on the hydraulic propeller, they also learned the intricacies of the Curtiss electric propeller which Rotol were making under licence and when the Rotol electric propeller came on the scene, this also became part of the course. Because one week's instruction was insufficient to ensure a thorough grounding in both the theoretical and practical aspects, the time allotted to each course was extended to a fortnight. Later, with the arrival of American aircraft in Britain, a necessity arose to differentiate between the Curtiss and Rotol propellers, but as the Americans did not use wood blades there was no need for them to be instructed on the Rotol types. Two new courses were introduced, and in the following months an increasing number of US personnel began to receive instruction. Service personnel from every Allied country attended the Rotol Instruction School and by 1943 some 6,000 had taken the written examination, which was the final hurdle.

All the Rotol departments had their part to play, but one of the most important and arguably the most interesting to the layman, was the Flight Test Department. Prior to 1939, all test flying was done at the parent companies airfields at Hucknall and Bristol, on aircraft loaned by the Air Ministry to Rolls-Royce and the Bristol Aeroplane Company. At the Board of Director's meeting in May 1938, it had been reported by Roy Fedden that some 500 hours flying had been achieved at Bristol on the Mercury-engined Hart, and 450 hours on the Pegasus-engined Vildebeest, both using Rotol hydraulic airscrews. The picture at Rolls-Royce was not so bright, Hives reporting that, 'Hucknall had testing facilities but no Rotol airscrews, except those manufactured by Rolls-Royce, and although these were not of the type being put into production by Rotol Airscrews at the present time, they were flying them at Hucknall whenever possible.' This rather unsatisfactory state of affairs continued for some months, until a separate department to control test flying was set up.

The Flight Test Department at Rotol was formed in the early part of 1939 under the overall control of R.S. Stammers the Rotol General Manager, but in 1944 Coverley appointed E.G. Fairhurst to the post of Chief Flight Test Engineer. The Chief Ground Engineer responsible for aircraft maintenance, services and ground staff was C.W. (Ben) Harris, an ex-RAF man who, according to those who worked for him, 'looked after his troops' and maintained a careful record of all the Ministry equipment on loan to the department. Any deficits would have to be made good by the firm, so we may be sure that Ben kept a close watch on such items. The main duties of the department were development testing of Rotol equipment fitted to single and multi-engined aircraft, which often resulted in suggestions for improving the performance of the constant-speed propellers, and alterations and modifications to the other Rotol products as the war progressed.

One of the early arrivals was Ted Edwards, who came to Rotol in July 1939 after serving 13 years with the Royal Air Force, followed by a six-month term with Air Service Training (AST) at Hamble as a ground engineer. He answered an advert for a ground engineer with Rotol and was offered a job with junior staff status at £6.10s (£6.50) a week, after being interviewed by Dick Morris, the Chief Test Pilot of the new department. The other pilot was Bryan Greensted, who did the lion's share of the test flying. Ted had known both pilots when they were all with AST at Hamble.

When Ted arrived on the scene, Rotol had two aircraft, a Vickers Wellesley, which had been taken on charge on April 17th 1939, and a Gloster Gauntlet. A third aircraft, a Bristol Blenheim MkI was added on September 1st. The fleet was increased by the end of the year with the addition of the first production Spitfire MkI serial K9787 and a Gladiator. Most of these and subsequent aircraft used by FTD were only retained for a short period while the test flying took place, although there were exceptions. Ministry of Supply contracts were often experimental, requiring extended periods of testing on military aircraft, and as a result those suitable for general testing were held by the department for several years and used on a series of contracts. A typical example of this was Wellington III serial BK537 which was operated for eight years, long after it would have been written off in RAF service. When its last contract was completed in 1952 this was what happened and it was bulldozed into scrap metal before return to them.

Test flying of the Wellesley commenced on June 6th 1939. Initially it was fitted with a De Havilland two-pitch controllable airscrew, with Schwarz wooden blades, but on July 17th the aircraft flew again, after the reduction gear from the Pegasus X engine had been removed and modified to take a Rotol electric constant-speed airscrew. One hundred hours flying with this aircraft was completed on December 20th 1939.

With the situation rapidly deteriorating in Europe, the call-up of reservists was announced on September 1st and Ted Edwards, being in the Class E reserve, duly reported for duty at Gosport. He had no sooner arrived when the authorities realised that his job at Rotol was far more important, and accordingly he returned on September 3rd 1939.

The Flight Test Department operated initially from a black corrugated hangar on the aerodrome, directly opposite the Rotol 'White Block', but a new building was clearly needed. A hangar and offices were constructed by the Derby firm of Ford and Weston, some 400 yards along the road towards Cheltenham – currently known as Dowty No 25 site – and in December 1939 the department moved into its new premises. Further aircraft arrived as the tempo increased, among them a Gladiator, a Battle and a Whitley.

Before the outbreak of war, Rotol had been in contact with some potential overseas customers, including the Fokker Aircraft Company in Holland, who had discussed an order in February 1938 for between thirty and forty airscrews. Because Rotol had not yet appointed foreign agents to act for them, the French Dewoitine company had also ordered airscrews through Rolls-Royce. In the summer of 1939, negotiations had started with the Belgian Air Ministry, for the supply of an airscrew for a Hawker Hurricane and if this was successful it would be followed by an order for a further eighty. The General Manager reported to the Directors at their meeting in July 1939, that a Hurricane fighter fitted with a Rotol airscrew had been sent to Iran by the Air Ministry. Even after the war began, Rotol were not relying solely on the orders from the Air Ministry, as foreign business continued to be pursued by Stammers and Newport, who made a visit to Holland and Belgium, in the early months of 1940, just a few weeks before Hitler's armies rolled into the Low Countries.

A month after the declaration of war, Hives reported that the Air Ministry had given permission to send 300 Rolls-Royce Merlin engines to France, all of them to be fitted to Dewoitine aircraft. The Board were disappointed to learn that the airscrews were to be supplied by De Havilland, despite the fact that Rotol had provided the prototype. In this connection, it was noted that a Service Representative was being sent to Toulouse, to supervise the installation of the Rotol propeller on the prototype.

The Service Representative in question was D.F. (Jock) Ancell, who admits that he arrived at Rotol by accident rather than design. He was motoring back to Coventry one day, along the Cheltenham Road, when he saw another car turn off just in front of him and enter the Rotol car park. Driving the car was a man he knew from earlier days in the car industry, Bert Jelley, and on impulse he followed him and asked what he was doing in these parts. Bert showed him around the factory and pointed out that with a war looming up there might be more of a future in the aircraft, rather than the car industry where Jock was currently employed. Jock thought about it and a short time later had an interview with Messrs Luby and Newport and agreed to join the firm on a three month trial basis in May 1938. His trial period turned out to be rather longer, as Jock stayed with the company until he retired in 1968. Within a short time he found himself fitting Rotol airscrews on a variety of aircraft, including the early Spitfire and Hurricane. The prototype and initial production versions of both these famous aircraft started life with fixed pitch wooden propellers, before graduating to variable pitch. At the end of August 1939, Jock moved North to fit the new airscrews to No 10 squadron Whitleys at Dishforth in Yorkshire, where the RAF fitters were hurriedly modifying their bombers to enable them to drop leaflets over Germany – the Government were anxious not to provoke retaliation from the enemy by dropping bombs on Germany at this period – but Jock managed to make a hurried trip home to be with his wife, who bore him a daughter on the day the British Government declared war on Germany, September 3rd 1939.

His next move was to take a cross-channel boat to France, to undertake the fitment of a Rotol airscrew on the prototype of a Merlin-engined Dewoitine fighter at Toulouse. The tests were successfully completed and Jock then went to another airfield near Paris, where a twin-engined French light bomber, the Amiot 356, was having its radial engines changed for Rolls-Royce Merlins and a pair of Rotol airscrews. Jock felt that the propeller controls needed to be alongside the pilot rather than behind him, and this delayed the test flight for a while. Eventually, at around two o'clock in the afternoon, the test pilot took the machine into the air and an anxious wait then followed before the sounds of the Merlin engines were heard approaching the airfield. After the Amiot had landed, the test pilot was full of praise for both the 'Magnifique' engines and the Rotol airscrews.

This was during the period which has become known as 'the phoney war', but a dramatic change took place on April 9th 1940, when Hitler unleashed his armies against Denmark and Norway, to be followed one month later by an attack on the Low Countries. On May 10th, Winston Churchill became Prime Minister, replacing Chamberlain, with a new Cabinet, including Lord Beaverbrook as Minister of Aircraft Production. In spite of the desperate opposition of the British and French armies aided by units of the Royal Air Force, the German blitzkrieg continued, and at the end of May the evacuation from Dunkirk began.

Jock Ancell was still in France, in company with the Rolls-Royce service personnel. These latter individuals had decided to take the three Rolls-Royce demonstration cars from the show rooms in Paris and make their way to Bordeaux, hoping to get on a ship back to England. Jock had other ideas. His plan was to hitch a lift home from the airfield at Le Bourget, if that were possible. He said goodbye to the Rolls-Royce crew (who probably thought him slightly mad) and, despite the problems from the German bombing and the French roads being choked with fleeing refugees, got a lift to the airfield. Le Bourget was practically deserted but he persuaded the British sergeant in charge (who just happened to be the brother of Tom Cruwys, another Rotol employee we will meet later in the narrative) to wangle him a seat on a Hudson aircraft which was taking the British Ambassador home to Britain. Due to the non-arrival of a French General there was one spare seat, so Jock Ancell finally and thankfully arrived back at Hendon in the late afternoon, to rejoin the Service Engineers and tell them of his adventures.

In charge of the Rotol Service Engineers Department was Lt Col J.D. Blyth OBE a well-known amateur archaeologist of some note, who contributed several theories on the Battle of Tewkesbury. He was always known as Colonel Blyth and there is an unconfirmed story that when he joined the Rotol Home Guard, he was addressed as Private Colonel Blyth.

For administration purposes the whole of the United Kingdom was divided into eleven areas by the Service Engineers Department, each with its own Area Supervisor and staff of engineers. One of the longest serving Area Supervisors at Rotol was Bill Burchill who specialised in electric propellers but had a wide knowledge of the whole range of Rotol products.

The Service Engineers maintained continual liaison with all the armed services, not only in Britain but also overseas. They were also stationed with various aircraft manufacturers, such as Vickers Armstrong, Blackburn, A.V. Roe etc all of whom provided an important feed-back to the Main Works on many aspects of the company products, sometimes resulting in a change to the design or manufacture. The incorporation of modifications to service equipment already 'in the field' frequently caused a tremendous amount of work and if this was too much for the RAF squadron fitters, the Rotol Service Department would be called on to undertake the work. According to the records, one such contract covered modifications to 1,100 propellers, occupying 7,000 man hours and over 17,000 miles travelling.

The Rotol Service Department engineers spent a great deal of their time during the war with the Halifax squadrons of RAF Bomber Command, mainly based in Yorkshire. Over 11,500 of the three-bladed and 3,600 of the four-bladed Rotol hydraulic airscrews were produced for the earlier versions of the Halifax. The rather more famous Lancaster bomber was fitted with the De Havilland Hydromatic product, except for a small number which had Rotol electric airscrews. Earlier in the war, before the four-engined heavies came on the scene, the brunt of the bombing campaign rested largely with the Vickers Armstrong Wellington, the Mk II version fitted with Rolls-Royce Merlin engines having

Rotol hydraulic airscrews and the Mk III with Bristol Hercules engines and Rotol electric airscrews.

When Jock Ancell returned to Rotol, he clashed with RHC at their first meeting – Jock was always somewhat outspoken, which would certainly not please Coverley – but after some initial difficulty he was appointed to the position of Repair Manager at the new facility in Gloucester. As already indicated, Coverley had few friends among those at Rotol, except for perhaps one, Denis Savory, and this is his story:

> My father started work at Rotol Airscrews some time in 1939, which meant that my family had to move from our Somerset village to come to live in Churchdown. At the time I was a fourteen year old apprentice jockey – I was only 4′6″ tall and weighed just over six stone. Apprentice jockeys were not much in demand in Churchdown and it was decided that I should become an apprentice at Rotol. Because of my size, there was some doubt of me becoming an apprentice, instead I had an interview with Stan Bannister for an office boy's job, where I faced a barrage of questions. I have to admit I liked him right away because he made it seem as if I was doing him a favour by accepting the position.
>
> The first day I started in the office of the Rotol Controller, the great man himself was absent but I was given a rundown of dozens of names to learn and different places to go. It was all too much for a little kid from the country and I made up my mind to get back home quick, but Mr Bannister gave me a straight talking to (and promised that the machine gunners on the roof had orders to shoot anyone trying to escape . . .) and I settled down to learning names and places. It was impressed on me in no uncertain fashion that RHC was Number One and the most important man in the whole factory. His orders took all priority; he could make fire come from his nostrils and eat little boys who did wrong. He was the equal of, if not more important than the Prime Minister!
>
> On the second day, I was sitting at my desk when the door behind me flew open with a crash, almost making me jump out of my chair with fright. No one else in the office looked up, which puzzled me for a moment. Standing in the doorway was a thick-set figure, studying me over the top of his spectacles. Secure in the knowledge that I was the office boy to the top man I returned his stare and told him that he'd better be careful how he opened doors in future because if RHC had witnessed this entrance he would not be pleased. At my outburst, the unknown man stood speechless for a moment, then he slowly hissed and said, 'And who the hell are you?' I stuck out my chin and replied that I was RHC's office boy. 'That's what you think – you're too bloody small' he bellowed and so saying he turned on his heel and marched off down the passage. The office came back to life, with much giggling and Mr Bannister shot out of the door in pursuit.
>
> Just then my buzzer, which was connected to RHC's office, started rattling like mad and I took off smartly and tapped on his door. Opening it, I was immediately impressed with the sumptuous green carpet, leather furniture and a magnificent desk. Leaning against it was the man I'd just spoken to and with a sinking heart I realised who he was. I thought my time had come but instead of giving me a right ticking off, he just smiled as he settled into his chair and once more gave me a look over his glasses. He gently chided me for my recent performance and from that moment we got along extremely well.

Denis Savory was eventually allowed to join the Apprentice Section about twelve months later, as apprentice No 50. He went on to complete almost fifty years with the company and had a number of further encounters with authority, but it is doubtful if any were of such proportions as the one in his first week. Denis spent a large proportion of those years in the Experimental Jig and Tool DO and was a great exponent of the drums, playing in most of the various Rotol bands.

The Rotol Experimental Department was built in the hard winter of 1939/40, largely due to Coverley's insistence, who recognised that the existing area in No 1 machine shop was far too small. It was occupied even before it had been completely finished. A tarpaulin at one end kept out the worst of the freezing wind, and braziers were used in the middle of the shop in a desperate but futile attempt to maintain a reasonable working temperature. The site chosen must have presented some problems for the builders, as it was on a pronounced slope and the floor level nearest to the Hatherley Brook was a considerable height above ground level. (There are references from various sources of components being deposited in the stream over the years by machinists on the nightshift who wished to hide their failures, but the story has proved impossible to verify.) The offices were situated on the balcony running along one side of the building, and housed the small Drawing and Planning Office, with offices for the secretary and manager. In the years that followed, the Experimental Shop was also used by the Research and Electrical Department, details of which will be recounted in a later chapter. At the time of writing these premises are 'the new technical block'.

N.S. Payne was transferred from Rolls-Royce at Derby in 1939, to take charge of the Experimental Drawing Office initially, but was soon given control of the whole of the Experimental Department. Charlie Beasant was the Shop Superintendent for many years. At the outset, the department worked exclusively on propellers but by the middle of the war it included all the other products with which the company were involved.

Charles Purnell started in the original shop, when it was in the corner of No 1 Machine Shop, and was always associated with the Experimental Department. He came to Rotol in May 1938, in company with a number of others from the Railway Works at Swindon, when that organisation laid off several hundred of its workers. One of his earliest memories is of the games of football played on the field where the car park now stands. In the lunch break, hordes of lads took part in 200-a-side matches, when you were fortunate if you ever got the chance to kick the ball! First employed as a universal miller, he soon found that he was called on to operate most of the other machines which began to fill up the shop, and as time went by his reputation grew. When something needed doing in a hurry, Charlie Purnell was the man to whom everyone turned.

At the end of September 1938 he watched as two German engineers installed a Forst Horizontal Broaching machine. This piece of equipment was to be one of the most useful ever bought by the company, broaching thousands of hub driving centres, day and night throughout the war, with very little maintenance. It continued in this capacity in the post war years and is still performing as well as ever, over fifty years later. The Berliner machines, purchased for £1,750 in 1937 and 1938, also continue to provide yeoman service, and the same story can probably be told about other machines, some of which are still in use by the present company.

Security was always of paramount importance during the war, and all factories producing weapons of war were subject to strict disciplines. Rotol and its dispersal factories were no exception. Not only were the entrances to the factory controlled by security guards, but the Experimental Department had its own security guard to ensure no one was admitted without proper authority. All personnel had to have a special badge in order to gain entrance, the possession of which lent an air of status among the messenger boys, whose daily tasks took them into this most secret place.

The Experimental Shop had a hard core of Left-Wing Socialists and was widely known throughout the factory, and indeed the district, as The Kremlin. There were also a number of card-carrying Communists until the Hungarian uprising in the 1950's which was brutally suppressed by the Communist authorities, when many tore up their cards. Those who worked in the offices or the shop floor were always in the forefront of changes in wages and conditions and had a reputation for being a tough bunch but with a close comradeship. Denis Savory was one of several who commented on the way the department always fielded a strong side in the inter-departmental matches, football, cricket, skittles – and frequently won them!

In the early 1960's, following the Dowty take-over, the original Experimental Shop was converted for use as Technical Offices, with the title of 'The New Technical Block'. (The Experimental Department was transferred to Arle Court for a short while, before returning to another building on the Staverton site.) The terminology 'New' has often been used to differentiate between an old building and a new one, and retains the title long after its 'newness' has ceased to be. This is the case of the 'New Inn' in Gloucester, despite it being several hundred years old, and although the New Technical Block is not quite that age, no doubt it too will continue to be known by this name for some time yet – probably until a new office block is built!

In the few short months since the declaration of war, the Rotol workforce had been transformed, certainly in terms of numbers, from the few at Llanthony Road to the many at Staverton and new names began to appear in the top echelons of management. Coverley was firmly in command and H.N. Charles had recently been appointed to the new post of Chief Engineer. Charles had been with the MG Car Company, responsible for racing and record breaking and according to those who knew him, was only too delighted to break into reminiscences at every opportunity. He made a number of important changes in the Design Office. L.G. Fairhurst became the Chief Designer, which now included all aspects of propeller design, the positions of Project Engineer and Blade Project Engineer being filled by 'Pop' Milner and Walter Brierley, respectively. Frank Blair went into LGF's old job, that of Chief Draughtsman, but after about four years he was forced by ill health to step aside, to be replaced by Eric Stockdale.

Bert Newport had been appointed Assistant Chief Engineer from 1st January 1940 and he and LGF were frequently required to visit other aircraft establishments all over the country, maintaining personal contact with senior personnel. One day they were performing their 'Mutt and Jeff act' (as LGF described it) after an urgent request from Sir Sidney Camm at his wartime habitation at

Claremount Palace, near Esher and were met by an irate Camm demanding to know why Rotol were offering him 'railway sleepers' for his new mark of Hurricane, instead of propeller blades. Apparently the airscrew Rotol had developed for the up-rated Rolls-Royce Merlin fitted to the Hurricane was actually reducing the performance instead of the expected increase. On subsequent enquiry it was found that there had been a miscalculation and manufacture of a new blade was immediately set in motion. It was due to this unfortunate episode that Charles made all the changes referred to above, in the technical set-up at Rotol. Although the new blades performed entirely satisfactorily, it was to be some time before complete confidence was restored in certain quarters.

The question of working hours at Rotol in the years before, during, and after the war has been the cause of some argument during the research for this book, as little written evidence is available and memories are not infallible. In 1937 the staff worked thirty seven and a half hours per week, as given in the letter of engagement quoted earlier for Jack Dancey. Works personnel are believed to have started at 7.30 am (possibly with a shorter lunch break), giving them a forty seven hour week, a figure which has been quoted in some documents. According to some sources, when the company first transferred to the Staverton site, the lunch break had to be extended to an hour and a half, because there was no canteen and the workers had to go home or visit 'Smoky Joe's', a well-known local café that was less than half a mile from the factory and certainly had a reputation for a quick snack. This extension to the lunch hour probably lasted for a very short while, as some form of canteen facility was quickly set up, even if it was only in a hut to start with.

Working hours changed completely with the coming of the war. A night shift was introduced, for men and women, and according to those who were there at the time, both day and night shifts lasted twelve hours, with the day shift starting at 7.30 am and finishing at 7.30 pm. Diana Murray Hill, in her book, *Ladies may now leave their machines*, (referred to in the chapter on 'Sources'), gives the starting time as 8.30 am which seems rather unlikely in wartime, with the lunch hour from 1.00–2.00 pm. There was a ten minute break in the morning when a lady with a tea trolley brought much needed refreshment and one of fifteen minutes at 4.15 pm when the girls went to the canteen, and presumably a similar arrangement operated for the night shift. Some confusion surrounds the working hours at weekends, but it seems that Saturday was a half day. D.M. Hill does not reveal what happened on Sunday although there is an indication that it too was a working day. There may well have been several changes to these arrangements as time went by, which may explain why people have different memories on this particular subject. In the absence of clear documentary evidence it is impossible to be completely positive on the question of working hours. (The reader may care to refer to the Appendices Section where further details of both working hours and rates of pay are given.) One other important factor needs stressing when we consider working hours. During the war, local time in Britain was fixed at an hour ahead of Greenwich Mean Time, which meant that there was British Double Summer Time from May to September and British Summer Time for the rest of the year. This arrangement gave daylight in the middle months of the year until almost midnight and was particularly useful for farmers, but conversely in the winter the workers had to face leaving home and returning in the dark, which some found rather depressing.

Except for occasional special reasons, which are mentioned in a later chapter, the office staff were not required to work a night shift, although their hours were nearly as long. At one period of the war, according to some informants, it was practice to always finish at 5.00 pm on Thursday and to have every third Sunday off. In addition, most workers, both in the offices and the factory, served in the Home Guard or ARP or fire-watching, which took up what remained of any spare time!

At the Board of Directors meeting held on February 15th 1940, a discussion concerning the question of Rotol manufacturing various auxiliaries was raised, following the appointment of D.R. Pobjoy to the new position of Auxiliaries Engineer. He had been an associate of Roy Fedden in the Cosmos Company, who were taken over by the Bristol Aeroplane Company in the 1920's, where sleeve valve engines were the order of the day. Although Rotol had no plans to become involved with sleeve valve engines, Pobjoy was offered the post of Auxiliaries Engineer at Rotol, a position he continued to occupy until the end of the war. From the minutes of the meeting it appears that Coverley may have overstepped his authority in this instance and the appointment was subject to the agreement of the Board. Coverley was questioned about the matter and was careful to explain that Pobjoy had severed his connection with aero engines, and was now solely employed on Rotol work. The Directors appeared to be satisfied with these explanations and in Hives view, 'it would be quite practical for the accessory gearboxes to be developed by Rotol, after being designed by the parent firm, sub-contracted by Rotol and brought back into the parent company's factory in peacetime – if they required the work'. When the Directors met again on April 18th, this discussion continued at some length and it was finally agreed, 'that Rotol should not embark in any way upon the manufacture of existing types of auxiliary gearboxes, already designed and developed by the parent firms.' This strict attitude was modified with the passage of time, but the Directors returned to the principle of auxiliary gearboxes being only a temporary arrangement on many occasions in the months that followed, indeed in 1945 the whole future of Rotol revolved around this point.

In May 1940, Coverley met officials at the Air Ministry to complain of a shortage of work and was informed that orders had been held back because Rotol had become too far ahead with its programme of airscrew production. This excess production concerned airscrews fitted with magnesium blades as Rotol were still waiting for decisions to be made by the Ministry concerning the wood blades and according to RHC 'he had thereupon scoured the Air Ministry for other types of work with which to fill the Rotol productive capacity'. One result of his endeavours was the setting up of an airscrew repair shop in the Flight Test Hangar and repair contracts were soon forthcoming.

At the next meeting of the Board of Directors, later in the month, Coverley put a report before them setting out a production scheme for the manufacture of auxiliaries, in order to ensure full employment. At this stage the discussion was of a general nature, but both Hives and Fedden once again stressed caution when it came to considering the manufacture of gearboxes, as both the parent companies would wish to take this work back into their own factories, at the conclusion of hostilities. In Hives view, the gearboxes that had already been designed by the parent companies, could perhaps be 'rationalised' and developed by Rotol on a sub-contract basis. For the present, Rotol would concentrate on three or four accessories of a fairly simple nature, and Hives and Fedden were left to make a decision as to which ones they might be. By July 1940, work had commenced on the chosen four: the Heywood Compressor, Heywood Regulator Valves, Bristol Beaufort Reduction Gearbox and Bristol Beaufort Oleo Legs with a number of others under consideration. At a later date, Fedden raised the question of Rotol continuing the manufacture of proprietary articles, as he did not consider it a satisfactory policy but the reply indicated that it was purely a move to get into the market, and to make the Ministry and industry aware of the Rotol name. During the next few months, additional orders were placed for Pesco Fuel Pumps and at the same time the Rotol Cabin Blower was being developed.

Providing auxiliary gearboxes to operate the various services in aircraft, had occupied the attentions of the aircraft engine manufacturers for some time. Previous practice had been to mount the accessories on the rear of the engine – or wherever it was most practical – but this had led to certain difficulties as the number of accessories required increased. The answer lay in a gearbox which was driven from the engine by a separate shaft, to drive whatever accessories the aircraft required. These might be either hydraulic or pneumatic pumps or electric generators, to provide power for operating the brakes, flaps, undercarriage, automatic pilot, etc, etc. The use of a separate gearbox, mounted on the aircraft structure, meant that engines could be changed without disturbing the multitude of pipes and connections between each of the accessories and the rest of the aircraft. This greatly altered the lives of hundreds of engine fitters in the RAF who appreciated the ease of maintenance this offered.

Several months were to elapse before contracts for Rotol Accessory Gearboxes materialised, in the meantime work progressed on the manufacture of the auxiliaries already mentioned. During the war, Rotol gearboxes fell broadly into two categories. The larger, known as the six-drive, carried six accessories and two generators, while the smaller three-drive had two accessories and two generators, with a majority of the internal components interchangeable. These Universal Gearboxes employed ball bearings throughout, but a later version was developed, in which most of the ball bearings were replaced by plain bearings, in the interests of conserving these precious items.

The first aircraft to be fitted with a Rotol Universal Gearbox was the Wellington Mk IV, to be followed by many others as the war progressed, such as the Bristol Beaufighter, Short Stirling, Handley Page Halifax and the Avro Lancaster. In addition, Rotol designed gearboxes for special applications, to meet the needs of particular aircraft. These were the Lightweight Four-drive, the High-speed Three-drive and Two-drive Generator Gearboxes. Altogether, Rotol produced a total of over 28,000 boxes by the end of the war, with the largest number being supplied for the Lancaster bomber. This figure included gearboxes being produced by the 'shadow' factories as well as Main Works. At the peak of production, some 2,000 were being manufactured each month, according to the records.

No 2 Machine Shop – the Auxiliary Gearbox Shop – was built in 1940, especially to manufacture this blossoming Rotol product, although it seems likely that before the shop became available some production work had already begun in No 1 machine shop. The Superintendent in No 2 Shop for most of the war was E.R. (Eric) Searles, who also had charge of the gearbox assembly shop. Eric joined the company in May 1942 and when the author interviewed him in 1989, he referred to the 350 men and women under his command as being among the very best, and he was proud to have been associated with them. Apart from being rather hard of hearing, he looked exactly the same as most of his colleagues would remember him, although he was in his 85th year. Eric was quite definitely a 'workaholic'. When he retired in 1971 as the Assistant Works Manager, he promptly got himself a liaison job with Messrs Poeton of Gloucester and went on to do some ten years with them. 'I wish I could get back to work', he told me, 'I hate sitting around on my backside.' We will meet Eric Searles again later in this narrative.

The Auxiliary Gearbox Shop had been designed to be completely self-contained (except for heat treatment) and an unusual feature was the inspection bay, which ran the full length of the shop. Between each two lines of machines was the inspection 'cage', with a wooden counter top. After completion of each operation, the machinist placed the component on the counter top for inspection and if anything needed rectification it was returned to the operator. A conveyor belt under the inspection bench took the accepted components to the intermediate stores. From these stores, the parts were fed to a moving assembly line, the speed of which could be controlled from the Superintendent's office. The oval track – with a number of fixtures or stations spaced four feet apart – moved at an average speed of about one foot per minute. As each empty fixture entered the stores, a main casing was bolted on and a complete set of parts loaded into the tray. Each operator assembled his or her part, as the fixture moved along the line until it was completed, removed, and sent to the rig testing department.

There was also an overhead conveyor system, which fed every department throughout the factory. It had one basic disadvantage; because of its accessibility components could easily disappear as it went from place to place, and maybe for this reason it was eventually dispensed with. In the gearbox shop it transported castings and raw materials to the machine shop and was the subject of a little tale, related by Phil Pryse. He was working on the night shift at the time, when he noticed a pay packet in one of the trays, gently circumnavigating the shop. He picked it up and found it belonged to Frank Humphries (the AEU convenor). It had obviously slipped out of his pocket. He went to see Frank on his machine and said, 'Frank, I'm a bit short tonight, can you lend me a quid till tomorrow?' Frank took his jacket from the peg and a look of horror came over his face as he realised his pay packet was missing – but the huge grin on Phil's countenance soon brought relief – together with a few well chosen Anglo-Saxon expletives.

In common with most other departments at Rotol, the gearbox shop had its share of memorable characters. Among them was a highly skilled setter by the name of Bert Gibbons. At the time of this incident, Bert was living in the Rotol hostel, and came to work on a ladies bicycle. It was his practice, particularly on night shift, to cycle into the workshop and park his bike at the end of his machine and then depart to the cloakroom. His workmates took advantage of this regular habit to spirit the bicycle away and on Bert's return he looked all over the shop for the missing bicycle but to no avail. Eventually someone took pity on him and pointed towards the roof, where Bert's bike, now completely covered in whitewash, hung suspended from one of the trusses.

The second tale, contributed by Dick Marfell, concerns the same Bert Gibbons, but this time refers to Bert's pronounced stammer. He was keen on the football pools and always had some tale of a near win to relate, when he turned up on Monday morning. One of the girls on this particular occasion, asked how he had fared on the previous Saturday. 'B . . . bb . . . bb . . .', stammered Bert, obviously somewhat agitated and shaking his head. 'Bolton?' she suggested. Bert continued to shake his head. 'Blackburn, Bury, Blackpool?' Still no response. 'Bradford, Barnsley?' At last Bert managed to regain the power of speech, 'B . . . bb . . . bb . . . Bloody Arsenal let me down!'

A name which will be recalled by designers and draughtsmen, not only at Rotol but also in many other engineering companies all over Britain, was employed at Rotol in one of the production offices in the early days. Frank Whiteley had written a number of books on gears and gear cutting, including a series of comprehensive booklets, entitled *Gear Data*, published by the Association of Engineering and Shipbuilding Draughtsmen, more commonly known as the AESD (the union which originally catered solely for draughtsmen). These booklets were widely used by everyone who was involved with this complex subject and for many years it was a standard work of reference, probably selling more copies than any other of the many booklets published by the AESD. Frank was also responsible for producing some of the company reference works on gear cutting, still in use today.

Gearbox manufacture at Rotol was one side of the picture, we need to look now at the part played by the design team. The Auxiliaries Design Office was formed in the early summer of 1940, about the time of the fall of France and the Dunkirk evacuation, with Pobjoy in charge. As we have seen, the origins of the Department lay in manufacturing products on a sub-contract basis from the parent companies, the first of these being the Heywood Air Compressor. Heywood was a French company and as the drawings were naturally in French, Vernon Christie and John Chilman were selected to carry out the necessary conversions from Metric to Imperial dimensions and to deal with all the subsequent queries from the machine shop. Although not perhaps the most auspicious start for a new department which eventually came to rival the propeller side of the Rotol business, it was at least a start!

The new design team followed this by starting work on a remotely driven gearbox, providing drives for the aircraft accessories for a new variant of the Vickers Wellington with Pratt and Whitney Twin Wasp engines, the Mk IV. Pobjoy had put the general layout together and this formed the basis for the finished design. Production of the Wellington gearbox reached nearly 500 and was the forerunner of thousands that came through the Rotol machine shops, in the years that followed.

Within a short time, the familiar problem of lack of space at Main Works meant the whole section moving to Imperial House in Cheltenham, the premises of E.J.

Burrows, a firm of printers, who continued to retain a small workforce in one of the offices. This was the nucleus of what rapidly grew into the Auxiliaries Drawing Office, with Douglas Pobjoy as Chief Engineer. His Chief Designer was George Bubb and Vernon Christie, first as Gearbox Section Leader and later, Chief Draughtsman. Two other section leaders were H.E. Storer (Engines) and A.H. (Arthur) Lane (Pumps). In his youth Arthur was looked on as one of the 'far-left', as were quite a number throughout the drawing offices at the time. He went on to complete the remainder of his working life with the company, where his extensive engineering knowledge was a great help to many younger colleagues.

• E.D. Wynn had charge of the development side, with N.H. Payne (later to become a director of Dowty Fuel Systems) as his assistant. From the outset the Auxiliaries DO operated entirely independently of the Rotol Design Office, giving rise to a certain amount of friction among some members of the senior management. Although acknowledged as a very capable engineer, Pobjoy had the reputation of flying into rages at the drop of a hat, which probably did not endear him to certain people and may also have contributed to his eventual dismissal. Further information concerning the activities of the Auxiliaries Design Office is related in a later chapter.

Dispersal to Gloucester and Cheltenham

Information about the many different Rotol offices and departments, both at Main Works and at what were known as 'Dispersed Areas', has not always been easy to come by. In some instances the author has been unable to contact ex-personnel to provide details and the written records are regrettably very sparse. Fortunately, prominent among the Rotol Departments that have not featured so far, there is one that not only has several references in *The Airscrew* but also individual contributions from those who worked there. It dealt with the repair of parts that were wrongly machined, worn or damaged.

A small Repair Section was in being from the start but in the early months of the war Rotol set up two specialised departments, following an instruction by the Air Ministry. This stated that all factories engaged in the manufacture of armaments, must ensure that all repairs were carried out at a distance of not less than five miles from the factory. Each of them were referred to, as was the custom, by their initials, RD (Repair Dept) and ARD (Auxiliaries Repair Dept). At the outset RD dealt with all repairs but eventually concentrated on airscrews. ARD was responsible for auxiliaries, such as governor units, feathering pumps and the like, but when gearboxes began to be made, they too went to ARD for repair.

In mid-1940, the Repair Department occupied an area of 14,500 sq ft in premises leased from the Gloucester Wagon Works, where it was anticipated that fifty airscrews a week would be handled. After a few months, it was realised that the existing space was insufficient to cater for the increasing number of airscrews being sent for repair and an additional 14,000 sq ft was allocated. In common with most of the Rotol dispersed sites, the factory was completely self-contained, with a machine shop, plating shop, first aid room and a canteen. All the expenditure on alterations to the Wagon Works, necessary to enable the new enterprise to flourish, was paid for by the Air Ministry.

The man in overall charge was B.W. Bleaney, general manager of Repairs, Services and Publicity. Before the war he had been a Rolls-Royce car salesman, and was often given the task of looking after important Rotol visitors, where his talents could be used to the full. His secretary, according to my informants, was from Bertram Mills Circus, as were a number of other employees, both at Repair and Main Works. At one period, the RD Works Manager was Bob Coverley, the son of the Rotol Controller.

The Repair Department had the responsibility of accepting airscrews damaged in training or by enemy action and completely refurbishing them back to their original standard, together with a brand-new log book. Repaired airscrews always carried the letter 'R' to denote this fact, in all other respects the difference between a brand-new and a repaired airscrew was impossible to detect. Airscrew blades probably accounted for most of the work in the early part of the war, and damage to wood blades was relatively easy to put right but both magnesium and hollow steel blades were also repaired, which was somewhat more difficult. By September 1940 the Repair Shop had already received 640 airscrews for treatment, 230 of which had been repaired and returned and by the end of the war thousands more had come through their hands.

The department had a mixed workforce, and in common with most engineering works in Britain during the war, women were in the majority. Photographs of the Repair Department published in *The Airscrew* for January 1944 show only a very few men and most of them are elderly, possibly some had even been brought back out of retirement.

A section of the Main Drawing Office controlled the salvage and repair of components and all the drawings had to be officially approved by a Ministry committee before they could be used by the shops. Although there were all manner of differing faults, some were easily put right. Damaged internal threads would be tapped oversize and fitted with a special wire insert and components incorrectly machined could often be plated to correct size. Parts which could not be reclaimed by 'salvage or repair schemes' would be scrapped but the company always went to great lengths to keep this 'scrap rate' as low as possible, particularly in wartime, for obvious reasons.

The Ministry of Aircraft Production operated a total of sixteen Repair Depots throughout the British Isles to deal with damaged aircraft equipment. These depots were responsible for receiving all damaged components and sending them back to the relevant parent company if repair was beyond their capabilities. Rotol had its own Repair Co-ordination Department, to liaise with this branch of MAP and the man in charge was Jack Staples. We will meet him again later in the story.

Larry Smith started as a centre lathe turner in No 1 Machine Shop at Rotol in 1940, on permanent night shift, but spent much of his time in the Repair Section, in the area which became universally known as the 'snake-pit'. (Possibly because it was overlooked by a balcony). He had memories of the buses used during the war being crammed full of people, standing on both the upper and lower decks. Sometimes they were so full that people had to get off and help to give them a push in order to get started and it was not unknown for the police to stop the bus and make people get off because they were so overloaded. Eileen Hunt recalled the fun they had when she and a number of others entered the front of the bus and moved through and out of the rear door and kept on coming back to the front, until the bus driver wondered how many more would get on!

Absenteeism was one of the problems that most companies had to face during the war and it was dealt with severely and could even result in a prison sentence for persistent offenders. An 'Absentee Committee' was formed at Rotol to consider cases and deal with them accordingly, but the composition of the committee and how many offenders came to their notice is not known. Diana Murray Hill makes reference in her book to girls who were frequently brought in front of the committee, for being absent without sufficient reason. An amusing aside to this serious subject came from Norman Grinnell, who was also a turner in the Repair Section. Norman recalled that one of the best of the centre lathe turners was a tall, gangling chap named Bill Porter, who had a pronounced

Gloucester accent. One day, Norman and Bill were talking of lateness when Bill finally brought their discussion to an end. 'I yunt never late', said Bill, 'If I be late I goes 'ome.'

It is in the Repair Department that we meet one of the Rotol men who could not possibly be excluded from a history of the company, W.A. Maclean. No one ever knew him as anything but 'Jock', by which it will be gathered that he was from North of the border. He joined Rotol on the fitting bench, in August 1938, at Llanthony Road, but when the transfer to Staverton took place he went into the area already described as 'the snake pit', in a small part of No 1 Machine Shop. This original repair facility continued at Main Works, even after the setting up of the larger organisation at Gloucester, presumably because there was sufficient work to keep them both going.

Jock Maclean became Shop Superintendent in the Repair Department, a post he held throughout the war and it was he who pointed out to the author that the Department handled a great deal of other vital war material, in addition to purely Rotol work, which even extended to repair of tanks for the army. He went on to eventually become the Works Manager at Rotol and British Messier towards the end of the 1950's, and many people will agree with the summing up by one of his compatriots, 'He was the best works manager that Rotol ever had, and generous to a fault.' Jock was a popular figure with almost everyone who knew him, both high and low and his rise through the ranks, while not unique, was certainly unusual and on the way he avoided one of the main pitfalls, that of estranging himself from his fellows in the process.

We must turn our attention back to the Flight Test Department, where changes were taking place following a tragic event which occurred at the end of 1939, when the Chief Test Pilot's flying career at Rotol was cut suddenly short. Dick Morris went to meet some friends in Oxford but in the black-out he missed his footing and fell heavily down some steps and died from a fractured skull. Bryan Greensted was promoted to Chief Test Pilot and an old friend, John Selby Hall, was selected as his assistant. Because of the colour of his hair, he was sometimes called 'Rusty' but most people knew him as Jack.

He was one of five sons, his younger brother Noel had been killed whilst flying in the Battle of Britain a few weeks earlier. Jack came to Rotol with quite a lot of flying experience. He had taken his 'A' licence at the ripe old age of 16 in 1925 but had spent many years in South Africa on a farm in Natal before returning to England in 1935 and qualifying as a pilot with Air Services Training at Hamble. From this time, until he joined Rotol in 1940, he flew with either Imperial Airways or Jersey Airways. He was flying with Imperial Airways as second pilot in the Short Empire flying boat *Courtier* when it crashed at Athens, from which he received minor injuries. Although he had no experience of test flying, under the tuition of Greensted he rapidly gained a reputation as a good test pilot.

With the heavy test programme being carried out by the Department, including new and improved propellers for Spitfires and Hurricanes, de-icing propellers on a Whitley and the first Curtiss electrically-operated propeller built under licence by Rotol, it was realised that a third pilot would be needed. In July 1941, Dennis Dickson, an ex-RAF squadron leader with a wealth of flying experience, joined the team.

He was commissioned in the RAF in 1927, a few months after his eighteenth birthday, and after gaining his 'wings' took an Instructors Course at the Central Flying School in 1930. Dickson transferred to the Reserve to join AST at Hamble – considered by many to be Britain's Air University – in 1936 and was promoted to Squadron Leader in 1938. At the outbreak of war he made repeated requests to join Fighter Command and instead received a posting to Bomber Command, eventually joining 110 Squadron at Wattisham, flying Blenheims on the early night operations. He stayed with Rotol until November 1943, when he moved to Rolls-Royce at Hucknall for engine development testing, following this with a three year period at Fairey Aviation until 1947.

This was an exceptionally busy time for the FTD and the test pilots worked a shift system, starting at early morning and often going on until dusk. In correspondence with the author, Dickson related that 'the work was interesting, indeed fascinating by reason of the variety of aircraft we were called upon to fly, from fighter to four-engine bombers. It was further enhanced by the accord and comradeship which existed between Bryan Greensted, Jack Hall and myself.'

An essential part of the airscrew was the spinner, streamlining the nose and housing the pitch changing mechanism. The first spinners fitted to Rotol airscrews were made by a firm from Warwick called Constant Speed Airscrews, although it is doubtful if they ever produced such a product as they lacked the engineering skills and machinery. They had negotiated a licence with the German firm of VDM to produce variable pitch airscrews before the outset of World War Two, but eventually settled for producing only the spinner. Their Chief Engineer, at least for a few years before coming to Rotol, was E.H. (Ted) Holden. Ted had served as a despatch rider with the Royal Flying Corps during the First World War, following this with a spell as an instructor for the RAF. By 1929 he was Assistant Chief Draughtsman at Ruston and Hornsby, moving once again, first to the Motor Division of Raleigh in Nottingham and then to Gloster Aircraft to join 'Pop' Milner in 1931. In 1934 he became Chief Propeller Designer for Armstrong Siddeley, transferring to Constant Speed Airscrews in 1936. By the end of the 1930's he began to realise that there was not much future in being associated with a German company and he rejoined 'Pop' Milner, now at Rotol, on Project Design. Ted became the head of the Technical Publications Department, when the need for comprehensive manuals explaining the workings and operation of complex propeller equipment became apparent.

Rotol's Chief Engineer at the time, H.N. Charles, designed a special locking feature for spinners, simpler than the VDM version, requiring only a screwdriver to lock or unlock. After passing the necesssary flight and development tests this was patented under his name, for which he is believed to have received the magnificent sum of one pound. From this time, Rotol began to manufacture its own spinners, and with the policy of dispersal firmly established, the production of spinners was moved to premises in Cheltenham High Street, formerly Steel's Garage, on September 11th 1940. (Occupied by C & A Stores at the time of writing).

As already explained, all the dispersed departments were self-contained wherever possible and the Spinner Factory was no exception. Bert Withey was the Superintendent, transferred, like almost everyone else, from the Main Works.

His secretary, who also looked after Welfare at the Spinner Factory, was Mrs Woodrow, who had joined Rotol on October 25th 1937. Entry to the premises was controlled by Sgt Randall, who was always careful to scrutinise all pass-holders. Just how many were employed in the Spinner Factory is not known but based on the photographs of the period, and estimates from those who worked there, it seems to have been around fifty altogether, with the usual high proportion of women.

The ground floor was entirely devoted to spinner assembly, the individual parts for each spinner being made either in one of the many sub-contract factories or at Main Works, and delivered to the Spinner Factory for assembly. This process involved drilling some 600 holes of various sizes, using portable drills because of the complex angles and contours and then fitting the rivets. A skilled riveter could handle about 3,000 per hour. The spinners were balanced and sprayed with 'Night Flying Black' and finally stencilled with the 'Lock and Unlock' notices. At its peak, the factory was handling about eight or nine separate designs of spinner, all for different aircraft and according to the records, a spinner was delivered to the Despatch Department every ten minutes of the day.

On the first floor at the rear were three smallish offices that were used – at least for a short time – by the Instruction School, under a Mr Shorter, aided by two instructors. The front office housed the Spinner DO and overlooked Cheltenham High Street and the Plough Hotel. In the early months of 1940, Jack Pickford, who had just started in the Drawing Office at Main Works, was directed to report to the Spinner DO where his Section Leader was Harry Horsfield. (When Harry passed away, before the end of the war, Tommy Robjohn replaced him.) On a sweltering hot day in the summer of 1940, with insufficient ventilation in the old building, the draughtsmen were moved to disrobe various items of clothing, Jack going further than most, being left in his underpants. Tom Godden passing by the office and seeing this almost naked draughtsman was moved to remonstrate with him, 'Get some more clothes on Jack, you never know who might turn up from the factory.' Half an hour later, Tom returned to find that Jack had done as requested, he had put on his tie . . . Jack spent 42 years with the company, almost the whole of the time being involved with the propellers, where, because of his extensive background knowledge, he became known as 'The Blade Expert'.

The topmost floor was shared between the Project Office under Ted Holden, with a small team, and the Detail Section where all the parts were drawn from the schemes that were prepared – at that time – in the DO at Main Works. The Section Leader was Tom Godden but when a new Project Design Office was created in the early part of the war, coinciding with the setting up of the Auxiliaries Section, Tom moved over to take charge, with the title of Chief Project Designer. In the reshuffle, a new title was given to the Project Designers to distinguish them from the senior draughtsmen and among this new elite band were Messrs Elmes, Scott, McNabb, Brooking, Holberton and Harvey, with Lloyd-Davis, Bishop and Locke as stressmen and technicians. Some names are probably missing from this list and not all those quoted were present in 1940.

Frank Morris was sent to the Spinner Factory soon after it started, as maintenance engineer. He not only looked after the machines but also doubled-up

"SUITABLE SONGS"
A NEWCOMER'S IMPRESSIONS AT
ROTOL
(ILLUSTRATED BY SONG TITLES)
"HAIL, SMILING MORN"
"PASSING BY"
"SMALL FRY"
"Somewhere a Voice is Calling!"
"Are you happy in your work?"
"Someday I'll find you!"
"Old Soldiers Never Die!"
"THERE'S A LONG, LONG TRAIL AWINDING."
"Pennies from heaven!"
"WHO'S TAKING YOU HOME TONIGHT?"
"GOD SEND YOU BACK TO ME!"
"Some day my PRINTS will come!"
• JOHN G. WALTER Accessories Drawing Office
Drawn 1941.

as First Aid Superintendent. He treated all the accidents on the spot whenever possible but serious cases had to be taken to Cheltenham General Hospital. During the war, great stress was laid on safety in the factories, particularly for women, who were urged to make sure their hair was tucked away under a cap, or a kind of net known as a 'Snood'. Frank recalled one girl who lost almost all her hair when she ignored these warnings, after catching it in a machine. There is an unusual and happy postscript to this tragic story. When her hair grew back some months later, she was surprised and delighted to discover that whereas previously it had been completely straight, it was now a mass of curls.

Turning now to another important part of any industrial undertaking, we must consider the role played by the body of men who are employed to look out for trouble. The Works Security Department's main task, particularly in wartime, has always been to ensure that Rotol did not fall victim to enemy agents and saboteurs, and even in peacetime vigilance has to be their watchword. In order to fulfill this function they control all works and office entrances, checking personnel, vehicles and goods, both incoming and outgoing.

During the war, the Security Department operated under the Emergency Powers (Defence) Act which gave far-ranging responsibilities, not only for the protection of the factory and employees at Main Works, but also at the dispersal factories and offices throughout the country. Acting as both the Works Security Officer and also the ARP Controller was A. Wilkinson. The uniformed Security Police under the control of the Works Police Superintendent, V.R. Hughes, comprised a large number of Patrol Men, whose duties encompassed Law and Order, Identification, Searching, Patrols, (including Guard Dog Patrol). They operated a three shifts per day system, every day of the year.

Covering the entrance to the 'White Block' and making certain that only those of sufficiently high authority would enter through this most holy of holy doors, was a commissionaire. For a large part of the time covered by this book, that office was held by Chief Commissionaire Taylor. He sat at a desk under the stairs (to begin with the stair railings were open but after some agitation from the girls who tripped up and down each day, they were panelled) and was responsible for the Visitor's Book, in addition to his other duties. There is a story concerning the Security Police, which may have involved Taylor as it occurred on his territory. The famous and well-known British Heavyweight Boxer, Tommy Farr, was passing the front office block in his car when he had a breakdown. Thinking he might be able to summon help, Tommy strolled up to the main entrance where he introduced himself and began to explain his plight to the policeman. The reply he received was to the effect that 'I don't care who you are – you aren't coming in here!'

The Works Engineer was B.S. Jelley, who had been with Rotol since July 1937. Bert Jelley had learned his trade in the motor cycle and motor car world, being Plant Engineer at both the Riley and Triumph works. His department had constructed or extensively altered premises at Gloucester, Cheltenham, Worcester, Manchester and other places in order to accommodate the Rotol dispersal units, as well as covering alterations to the main Rotol factory at Staverton.

The Rotol security gatehouse, or 'New Time-Recording Station', which was its official title, has gone down in history as one of Bert's failures, or at least one that he would have preferred to forget. It was erected in 1942 and still stands at the corner of the factory site at the time of writing and now houses the personnel office. Before it was built the police controlled the factory entrance from a hut each side with the gates between them. Bert decided that the time had come to introduce a stricter control over the comings and goings at Main Works, and to perhaps keep an eye on the clocking arrangements at the same time. Previously, it had been standard practice to use the clock at one's work place, merely showing a pass to the gate police on entering the factory. (Until the 1960's almost everyone had to clock in and out, the only exceptions being the senior staff.)

The building comprised a ground floor, with toilet facilities and offices for the Works Security Officer, Police Superintendent and others on the first floor but it was the layout of the ground floor which caused the problems about to be recounted. Basically, the interior was a bare hall containing racks of clockcards and several clocks. Employees, on starting or finishing their shift, entered the building through a double door at either end, to comply with the blackout regulations and were then channelled through one of eight lanes, segregated by metal barriers, before using the clock. After showing a special identity card to the police they were allowed to proceed. When the first day came to use the new system – all the clocks having been repositioned during the previous night – the nightshift started to clock out at the same time as the dayshift attempted to clock in. The staff began to arrive about an hour later, resulting in utter confusion with everyone fighting to get to the clocks and works employees still trying to get either out or in! Loud cries of 'Baa, baa' were heard coming from the building as the workers milled around within. It was mid-morning before order was restored and during the following night all the clocks were returned to their original positions.

From that date, the building was always known as 'Jelley's Folly', and some fifty years later, older employees still call it by that name. L.G. Fairhurst attempted to offer a crumb of comfort to Bert Jelley, by pointing out that he was at least in the unique position of having a building named after him, and how much he would have enjoyed the honour, but Bert refused to be consoled. Shortly after this episode it was decided to give an official name to all the roads on the site, possibly to avoid any further derogatory references of this nature. The gatehouse has been subject to many alterations over the intervening years and now bears little resemblance to the original plans, although some of the upstairs rooms may be unchanged.

One of the undying memories of the Second World War, thanks partly to the television programme 'Dad's Army', must be the citizen's army known as the Home Guard. Almost everyone who served with this fine body of men had a funny story to tell, usually involving long nights, itchy uniforms and general discomfort. A detachment of the Home Guard, originally called the Local Defence Volunteers (LDV) was formed at Rotol towards the end of May 1940 and the first LDV commander was Captain D.R. Pobjoy but pressure of work became too much for him and Major J.Y. Baldwin took over on 17th June 1940. Subsequent commanders were Captain Hayden and Captain Swann. Mr Stammers, writing in *The Airscrew*, described the beginning as 'full of enthusiasm but sadly deficient in equipment. Our particular section (Rotol) was perhaps more fortunate than some of the others in the area as we were able to get 80 rifles with 5 rounds of ammunition per rifle, through the assistance of the Chief Constable of Cheltenham.' After long hours in the factory and offices, it was hard, even with the extra pay of 3/- (15p) per week, to find the motivation needed to put in the required number of attendances on parade, which were in the order of two a week, and to learn the skills of foot and arms drill. This must have been particularly so in the later months of the war when the immediate threat of a German invasion had long since passed. Whilst sympathising with their plight, we should not forget that thousands of men and women serving in the armed forces throughout the world would have been only too pleased to have changed places with them!

Everyone who served in the Home Guard has some humourous incident to recall and nearly all of them tell of things going wrong in some fashion. This first story comes into that category and although it has not been verified it seems to have more than a grain of truth. It is reputed to have happened at the time of the fall of France, when an announcement was made over the radio that German troops had landed in Jersey during the evening. Someone misinterpreted this news and a strong contingent of the Rotol Home Guard set off for nearby Dursley! On arrival they were somewhat nonplussed to find the sleepy Cotswold town even more at peace with the world than usual.

Although it sounds more like the script for an Ealing comedy film, a second account also involved the Rotol Home Guard. In order to assist Rotol on a rather difficult propeller straingauge test, still in its infancy in Britain, two scientists from the Royal Aeronautical Establishment at Farnborough came to the factory. They arrived at 10.00 pm on the night before the test was scheduled to take place, so that the adhesive used to fix the gauges on the propeller would set hard overnight. The night commissionaire, not being privy to this information and ever concious of the threat from Fifth Columnists, locked them in a room adjacent to the main entrance hall, while he telephoned H.N. Charles, the Chief Engineer, who was also an officer in the Home Guard. Charles gave instructions to leave them there until he arrived, and then proceeded to round up an armed guard and also alerted the Chief Constable of Cheltenham. By the time they all turned up at Rotol, the supposed saboteurs were fast asleep in their chairs and were somewhat alarmed to be woken up, faced with a group of armed men pointing loaded rifles at them. Explanations and apologies followed, but the Chief Constable was not mightily pleased and he demanded that Coverley give a suitable admonishment to the luckless Charles the following day!

George Pettit was a member of the Home Guard, and remembered how they would be required to leave their machines once a week at 6.30 pm instead of 7.30 pm to report for duty. They wore their uniform underneath their overalls, which was most uncomfortable, and every month there would be night duty at the anti-aircraft site at Longlevens. After night duty, with only a limited amount of sleep, the men of the Home Guard would report back for work at 7.30 am to start another day shift, still in uniform.

The anti-aircraft site was at a farm in the vicinity of Nine Elms Road, Gloucester, opposite the greyhound racing track (demolished in the mid 1980's to make way for a housing estate) and in place of the conventional heavy gun more usually associated with Ack-Ack batteries, this one was equipped with rockets, laid out in 8 rows of 8 projectors, each firing two rockets, making a total of 128 in all. The rockets could be fired in a salvo of sixteen at a time, but so far as is known the Longlevens battery was never used against enemy aircraft. Ivor Brooking was one of those transferred to the rocket battery and after eight weeks of dry-run training went off to the firing range near Penarth, where the men could accustom themselves to firing this impressive and noisy weapon. Ivor recalled that 'when the first salvo was fired I saw more space between the feet of those Home Guard members present and the ground, than I ever imagined possible.'

Ron Brown began his working life at Rotol in 1939, as a joiner in the wood blade shop. Although he joined the Home Guard for a time, serving under Jack Shaw and Harry Greville, he was approached to transfer his allegiance to the National Fire Service. By this time he was with ARD in Gloucester and found it was a far easier life than in the Home Guard. In company with Jock Maclean and a couple of others, he spent some happy hours, training and demonstrating the rapid response of the men on the pumps. This was put to good use on more than one occasion when tackling the odd blaze in the paint shop, but he never had to deal with fires started by enemy action. Ron went on to serve the company until retirement in 1977 and will be best remembered as the Superintendent in the Plating Shop, and for his long association with the Rotol Old Time Dancing Club.

Jack Sullivan was a turner in No 1 Machine Shop in May 1940, under Charlie Homer. He had previously been a heavy goods driver, so he was completely unfamiliar with the machines which now confronted him, but undér the expert tuition of the friendly instructors, he quickly mastered the intricacies of fine limit boring and other complexities. In many ways his experiences during the wartime years were typical of the large number of employees from all walks of life and from far afield, who had been recruited by the company over the past few months. One problem resulting from this blending of skills from opposite ends of the country was the difficulty of communication, as the people from the North tried to understand those who hailed from Bristol and vice versa. Added to this was the almost incomprehensible dialect from the Forest of Dean dwellers . . .

All the newcomers came from varied backgrounds, a few had even worked at a local circus, as has already been noted. Under the Conscription Act, every able-bodied person was required to register for war service and women, as well as men, were required to 'do their bit' for the war effort. For some of these ladies in particular, it was an experience that changed their whole lives. They received training in various Government Centres all over the country before being directed to one of the factories engaged in producing weapons and munitions. As time went by, because of the lack of men, women undertook all manner of jobs that would have been considered far beyond their range in peacetime.

At Rotol, women underwent further training and a great many ended up in the Inspection Departments, where there was a particular shortage of skilled men. In the Machine Shops a system was developed to reduce the engineering knowledge and skill to a minimum, coupled with more special training, thus enabling a high proportion of women to undertake the role of inspector. The ratio was quoted as being as high as five women to one man at the peak, and was acclaimed a great success. One means of improving the expertise of the men and women was by means of the use of instructional films, and Rotol are believed to have produced some of those that were shown in the Training School. The man responsible is thought to have been A.N. Redmond, whose name is more often associated with the ARD Concert Party, where he entertained his audience with a variety of illusions.

Wartime Rotol was a very different place from what had gone before, both outside and inside. A sense of urgency was present and there was a comradeship among the workers, which, sadly, could probably only flourish in time of war. Great efforts had been made to merge the factory into the countryside, roofs and walls were painted with irregular areas of green and black, still faintly visible in places, even today. Suspended over the bus and car park by thick wooden poles, was a huge camouflage net, strong enough for a man to walk on. After it had snowed, it was like walking through a strange grotto, but with the thaw came the constant and unpleasant drip of the melting snow. From the air, photographs reveal not only the extent of the camouflage but also the large number of huts on the site, some of which were used as storage sheds and others for Home Guard duties. Those on the South Works site survived into the 1970's and were used by some of the RAF units who were stationed on the airfield.

To protect the workers from air raids, shelters were built in several parts of the factory and from the original plans held in the Works Engineer's Department it is possible to see exactly how many there were. In front of the main offices, running alongside the main road, the plans show a total of 18 shelters built below ground, and what appear to be a further 5 in Down Hatherley Lane. These are in addition to surface shelters attached to the outside walls of each of the machine shops. A large underground shelter is said to have existed in the field alongside No 1 shop at the start of the war but when No 2 shop was built it was replaced by a surface shelter, situated in the roadway between the two machine shops. The surface shelters considerably reduced the width of the roadways between the machine shops, as can be seen on aerial photographs. Although there are many romantic tales about the Rotol air raid shelters, they tend to be rather repetitive and few can be printed! The following story, which has been slightly edited, may be considered typical.

Charlie Homer was conducting a Ministry Official on a tour of the factory, and inadvertantly interrupted a couple in one of the shelters. Charlie hurriedly stepped back and shut the door, somewhat embarrassed, and turned to the official saying, 'Sorry about that, I will have to tighten up on discipline', but the official replied, 'Don't worry too much, it's the most activity I've seen at Rotol since I got here.'

All the standard wartime blackout and blast precautions against bomb damage were applied throughout the works. Double blast walls were built in the bus park, (one of which still stands at the time of writing) and they were also put up in the machine shops. All windows were protected with strips of cloth, stuck on to the panes as a precaution against flying glass and splinters. Doorways were constructed so that one had to pass through an inner door before going outside, to prevent any stray light from showing at night. In addition, all windows were painted black, to comply with the blackout requirements, which resulted in the factory being lit continuously by mercury-vapour lamps, giving everything and everybody a greenish-white appearance. During the day it was possible to open windows and parts of the roof for ventilation, but at night time a kind of haze, from the cigarette smoke and machine oil vapour, filled the machine shops.

The problem of ventilation was not confined solely to the machine shops. At the beginning of the war, the company had an urgent contract from Vickers Armstrong to make part of an undercarriage for their Wellington bomber, and a few personnel from the Drawing Office were asked to work a night shift. John Boast was one who volunteered and he recalled the atmosphere from the cigarette and pipe smoke was so bad in the blacked-out office that they had to switch off the lights and open the windows at intervals, to let the 'fug' disperse. It is probably difficult for youngsters to fully appreciate the extent of smoking which was a normal part of everyone's existence fifty years ago. Wartime life was remembered as a time of shortages not only of food but – most importantly for the great mass of the adult population – also of tobacco. This is most noticeable in old films of the period, when anyone who is not smoking seems somewhat out of place!

Still to be seen around the factory and aerodrome are the pill boxes, one of the few remaining features of wartime Britain, that are gradually being dismantled. They would have been used by units of the regular army or the Home Guard if the German invasion, which had been projected in the autumn of 1940 had taken place. The German Wehrmacht would probably have proved a difficult adversary, based on the success they enjoyed in Belgium and France, but fortunately the fighter planes and pilots of the RAF defeated the Luftwaffe, and without mastery of the sky, Hitler wisely decided to leave Great Britain alone while he turned his attention to the USSR.

One of the common fears of the early war years, particularly following the fall of France in May 1940, was that of Nazi paratroopers attempting to take-over key factories as a forerunner to a German invasion. This threat was taken seriously at Rotol, where a special anti-parachutist patrol was instituted by Stammers. Two of its members were Eric Stockdale and Charlie Palmer from the DO who were issued with rifles and a ladder to enable them to climb over the many fences around the factory, without going through the gates. Eric recalled that he had cause to complain when Charlie was somewhat careless with his rifle as he followed Eric up the ladder one particularly dark night. This story, with two men groping around in the black-out, armed with ladders, rifles and possibly a dim torch, conjures up a picture of the great comedians Laurel and Hardy at their most sublime but Eric solemnly assured me that it actually happened. The sleeping arrangements for the off-duty patrol also lacked some creature comforts, being merely a mattress on the floor of the drawing office.

The possibility of an invasion was probably taken more seriously in the South and South-East than in the West Country which was likely to be more remote from the initial landings but precautions still had to be made. It was in connection with the threat of invasion that the following story was repeated to the author on more than one occasion, although the details frequently varied. To inform the workers of the imminent arrival of Nazi paratroopers, and remind them of the action they should take, a recorded message would be transmitted over the tannoy system. 'The invasion is now in progress. Will Home Guard members report for duty immediately. Workers who live nearby should return home and all others take to the open fields.' (The words may not be exact but the gist is there.) On hearing this message, the machine operators were expected to remove a vital part of their machine to render it inoperative and disperse as ordered.

On this particular night, someone in the Control Room was demonstrating the system to a colleague, apparently not aware that the tannoy was 'live'. Suddenly, from the loudspeakers came the dreaded message and one can imagine the effect this must have had on all who heard it. The workers sprang into action and took themselves off with all possible speed and although the order must have been countermanded almost at once, it was too late for some, who had made straight for the nearest public house and were quite happy to await further events. It took the Works Security Police a few hours (some people suggested 'days'!) to find everyone who had left so promptly. Afterwards it was probably looked on as a 'a bit of a lark' but at the time many people confessed to being terrified when they thought that the German troops had landed.

No doubt there were some repercussions and perhaps even disciplinary action taken but the outcome was that the idea of using the recorded message for instant evacuation was quietly dropped. Once again, Murphy's law had been proven, 'If it is possible that something can go wrong, sooner or later it will.'

Rotol and its environments, in common with many other wartime factories, was equipped with a smoke screen in case of possible German air attacks. 'Hal' Reynolds had charge, and supplied the following details of its workings. The factory and a large part of the immediate area in the vicinity was surrounded with smoke pots, (code named 'Little Red Babies') operated by army personnel who were billeted in Shurdington. Whenever there was a likelihood of enemy air attack, Hal would receive a coded message on the telephone from the War Office in Whitehall and he then had to ensure that the up-wind smoke pots were lit. Weather conditions were not always favourable for this operation but if successful, a dense cloud of oily black smoke would cover the factory and much of the surrounding countryside. The smoke screen could last for several hours and the smell lingered long after the smoke had dispersed. Everyone hated it, nor was it too popular with the local housewives. How many times it was used during the war is a matter of conjecture, but the smoke screen left an indelible impression on the memories of many people, making it appear to have been used far more than it really was. It seems unlikely to have been needed much after 1941, as the Luftwaffe made only rare appearances in the West of England after this date.

It was certainly in operation on the night described by George Pettitt. 'One moonlit night in December 1940, the night shift had just arrived and soon afterwards the sirens sounded. We all trooped down to the shelters, with the Rotol smoke screen much in evidence, for what we thought would be a short time and started singing, but as the night wore on we relapsed into a miserable silence, with the bitter cold and the night's bonus gone to pot.' This was possibly Cheltenham's heaviest air raid of the war, with several houses in the area of the Lower High Street demolished, some people were killed and many more injured.

One other employee who had cause to remember that night was Reg Russell. He had been working at GAC until he and about twenty others were directed by the Ministry of Labour to report for work at Rotol. Part of his job was to look after one of the outlying blade stores, at Peacock's shop (behind Steel's garage in the High Street) where the blades were standing on end, awaiting transport to Main Works for assembly. Following the bombing raid on Cheltenham, he went into the store next morning, to find that every one of the blades had fallen over, due to the earth tremors caused by the explosions. Fortunately, they only received superficial damage which was easily repaired, but thereafter the blades were always stored lying down.

Air Raid Sirens were situated in several parts of the works to give warning of the approach of enemy aircraft and the responsiblity for sounding the siren rested with the Works Police in company with the aircraft spotters on the roof. This was to ensure that any interruption from the enemy would result in only a minimal loss of production. Inside the factory, where the noise of machines might cover the wail of the siren, a system of coloured lights was used. A blue light meant 'Enemy aircraft approaching', a red light, 'Make your way to the shelter' and a white light 'All Clear'. These same coloured lights were also used to call key personnel to the telephone, different combinations being used for different people. Although these coloured lights were fitted in the Rotol offices, the tannoy system was excluded for obvious reasons. The BBC radio programme, 'Music while you work' was broadcast over the loudspeakers for half an hour during the morning and afternoon and was always a great favourite, particularly among the ladies, and it was not unknown for the office girls to find some reason for going into the machine shops when this was being broadcast.

Lord Beaverbrook, the Minister of Aircraft Production, sent the following stirring words in a telegram to the firm on July 5th 1940. 'Will you please tell your staff that I will welcome and rejoice in a decision by them to stay at their posts if the sirens sound and until the danger of bombing attacks is imminent. Stop. Such a demonstration of courage and devotion would in large measure defeat the enemy's purpose of damaging our production for defence. Stop. If every factory is a citadel of resolute men (Beaverbrook had overlooked the fact that a large number of women were also playing their part – author) sustaining by their labours the pilots and crews who battle for us in the skies, Britain will hurl back the onslaught of our enemies and save the world for freedom.'

In the event no bombs actually fell on Main Works but some RAF men were killed from a bombing raid on the nearby aerodrome and the Rotol Flight Test sheds. This may have been the same occasion witnessed by John Chilman and a number of others in the drawing office, when a lone German aircraft came across the aerodrome at a very low height and John broke all records to get down the stairs and into the nearest shelter. Another Rotol eye-witness, Dennis Savory, was in the Apprentice School and saw the aircraft bank and drop its bombs near the Hare and Hounds public house, raking the area with machine gun fire. Neither the light machine guns to be used for factory defence, nor the rocket battery at nearby Longlevens were fired on this occasion, but folk legends maintain that the German raider was subsequently shot down near the South Coast. Whether or not Rotol was protected by barrage balloons has been a matter of argument but there is agreement that they – and searchlight batteries – were present in the locality.

Before leaving this chapter, we must look at the part which Rotol played in one of the events that is generally regarded as a turning point in the Second World War.

R.H. Coverley devoted himself tirelessly, not only to his job with Rotol but also to another, which he was offered in the early summer of 1940. Soon after the fall of France he was summoned to a meeting in London with Lord Beaverbrook, who had been chosen by Churchill to become the Minister for Aircraft Production. This was a brand new ministry, formed on May 14th 1940 and separate from the existing Ministry of Supply. Beaverbrook used businessmen from all over British industry to sort out production bottlenecks and Coverley was given the position of Aero Engine and Airscrew Advisor on his staff. Thereafter, Coverley divided his time between Rotol and visits to aircraft factories in connection with the job from the MAP although his involvement at Staverton inevitably lessened as the work for the Ministry took precedence. Exactly what this job entailed is not known for certain but his secretary Miss Jackson maintains that it was largely due to Coverley's efforts that Rotol stockpiled a large quantity of wood blades that were to be desperately needed in the next few weeks. After a year in office, Beaverbrook was replaced by Col J.T.C. Moore-Brabazon, during which time he had transformed the British Aircraft Industry. Although no records have been found to show that Beaverbrook ever visited Rotol it is known that his successors, first Col J.J. Llewllin and later Sir Stafford Cripps, came to the factory on more than one occasion.

The Battle of Britain, one of the decisive battles of the Second World War, was fought out in the skies over Southern England during the summer of 1940. RAF Fighter Command which faced the Luftwaffe, was mainly equipped with Spitfires and Hurricanes, the latter being by far the most numerous. At the start of the battle, there were 19 squadrons of Spitfires and 38 of Hurricanes, but Spitfire deliveries soon improved this ratio.

A total of 1,531 Spitfire Mk I aircraft were built, the first 76 with fixed pitch, two-blade wooden airscrews, the remainder with De Havilland two-position three-blade metal airscrews and a similar story could be related for the propellers fitted to the Hurricane Mk I. To improve performance, most of these aircraft had their De Havilland two-position airscrews changed to constant-speed types, in a frantic conversion by De Havilland service engineers and RAF fitters which only began at the end of June 1940, and by August 15th a total of 1,051 aircraft had been modified.

Rotol began production of constant-speed airscrews with a pitch range of 35° in early 1940 and documents give the following total numbers of airscrews supplied for the two fighters: Hurricane Mk I – 2,394, Spitfire Mk I – 122, Spitfire Mk II – 1,193. From these figures it can be seen that only a small number of the Mk I Spitfires had Rotol CS airscrews, with the majority being fitted with the De Havilland version. However, large numbers of Rotol airscrews were fitted to both the Hurricane Mk I and Spitfire Mk II which seems to tally with the figures quoted in the official reference works, although it seems to be generally agreed that few Spitfire Mk II's took part in the battle as the first deliveries did not begin until late summer.

Whatever the facts of the matter are, by August 8th 1940 the majority of these front-line fighter aircraft had either Rotol or De Havilland constant-speed airscrews, indeed the Spitfire's coming off the assembly lines since the beginning of July were all fitted with them. As the Messerschmitt Me109 fighters of the Luftwaffe were already equipped with the VDM (German) constant-speed airscrews, the RAF could now engage them on something like equal terms. The constant-speed airscrew improved take-off and landing performance and also the aircraft's climb rate, and made control easier for the young and increasingly low-hour RAF pilots who had to combat the German attacks during those crucial weeks.

As the battle developed and the RAF losses mounted, the Civilian Repair Organisation, under Beaverbrook, produced a minor miracle by taking the aircraft that had been struck off strength because they were too badly damaged to be repaired locally and returned them to the squadrons as good as new, with repaired or brand-new airscrews as necessary. According to 'Jock' Ancell, the responsibility for looking after the Rotol propellers fitted to the Spitfires and Hurricanes of the Royal Air Force in the South-East rested with two Rotol Service Engineers, Jack Witts and Cyril Morton. Despite their losses during this period, RAF Fighter Command had more aircraft at the end of the battle than when it began, partly due to the efforts of the CRO. With the coming of autumn, the Luftwaffe turned its attentions from the destruction of the RAF and its aerodromes and began the bombing campaign of London and other big cities.

It had been a testing time for the British aviation industry and both Rotol and De Havilland contributed to the eventual victory. Some historians maintain that although it was the shortage of pilots that was to give most concern, the fitting of the constant-speed airscrew played a crucial part in the Battle of Britain. Although arguments still continue over the tactics of both sides during this time, what is certain is that had the RAF been defeated the German plans for the invasion of Britain would surely have followed, although whether that would have been successful is also hotly debated.

Some time later, the Air Ministry introduced the practice of sending RAF aircrew who used the equipment, be it aircraft, propellers or landing gear, to the firms concerned, in order to encourage the workers who were producing the goods for the war effort. Rotol is known to have had visits from several, including Flight Lieutenant 'Sailor' Malan and Flight Lieutenant Gleed, both of whom fought in the Battle of Britain as Spitfire and Hurricane fighter pilots. Some months later, Rotol presented a number of tankards – or 'pots' – to certain RAF squadrons, inscribed with the Rotol 'wings' in blue enamel and the total 'kills' to date, but whether any still survive in present-day RAF messes is not known.

The Mid-War Years

By 1941, production of Rotol airscrews was setting new records, with orders amounting to 19,200 of which 6,000 had been completed by the end of 1940, but many problems still awaited a solution. At the Board Meeting in January 1941, the Directors were informed that the MAP airscrew programme had not been finalised and some concern was expressed regarding the blade situation, as the wood blades were in difficulties. The problem was caused by Jablo Propellers of Stockton, who were having trouble with the glueing process and consequently falling behind with deliveries of the blocks and blades. The situation must have been desperate as the possibility of reverting to magnesium was considered. It was reported at the same meeting that Rotol might be required to produce 500 airscrews per week by the end of the year and the Directors believed this would certainly need the building of a shadow factory, in order to achieve such a figure.

Wood blades were being produced from both Jablo and Weybridge blocks up to this time but a third supplier was in the offing. Hydulignum was the trade name for the compressed wood which had been developed by the firm of Hordern-Richmond, a small pre-war aircraft manufacturer, who also made fixed pitch wooden propellers. They had several factories in the area around Chesham, in Buckinghamshire, which was the centre of the wood-working industry at that time. Hydulignum wood blocks comprised a constant number of birch veneers throughout the length of the boards, achieving the required increased root density by the application of vertical compression over the board length and additional lateral compression at the root end.

Rotol first ordered 600 Hydulignum blades on a trial basis in April 1941, on the understanding that further orders would follow if they were successful. At the same time, the Directors began to consider taking a controlling share interest in Hordern-Richmond, reasoning that it would make good business sense to control a company that was likely to be in the forefront of wood-blade manufacture in the post-war years. Although no action was taken at this time, in 1943 the idea was to be resurrected once again.

The Hordern-Richmond company were honoured by a visit from King George VI and Queen Elizabeth on June 27th 1941 and spent the afternoon in the Haddenham Works, being shown the Hydulignum blade processes. Photographs in *The Airscrew* show a youthful Queen, wearing one of her customary hats, accompanied by the King in Air Force uniform.

The familiar problem of lack of space to cater with the flood of work in the Repair Department at the Wagon Works in Gloucester continued to give concern as the war progressed. To provide more room, the company requisitioned Robert Brothers' factory in Upton Street, Gloucester, the pre-war producers of Glevum Games. This became ARD (Auxiliaries Repair Dept) but was usually referred to as 'the toy factory' by those who worked there. When Ron Brown arrived as part of the workforce soon after it opened, he was amazed to see that it still had boxes of bricks, snakes and ladders, jigsaws and toys of every description all over the place, and some time elapsed before they were cleared away. In overall charge was B.W. Bleaney, who already ran the other Gloucester repair facility, with Jack French as the Assistant Chief Inspector in charge of inspection at both RD and ARD Jack had been the second turner to be employed by Rotol at Llanthony Road in 1937, the first being Ted Marshall.

Airscrews and Governor Units received for repair were stripped, inspected and a report compiled, stating whether each part was serviceable, capable of being salvaged, or reduced to scrap. After reconditioning, the parts were again inspected before issue to the assembly shop, where the inspection and testing procedure were the same as for new production. To indicate the importance of the Repair Department, figures quoted when the war ended gave a total of no less than 50,000 airscrews and 31,000 governor units as having been put back into service, and the repair factories dealt with many other Rotol products besides these.

Wally Webb went to the Repair Department in the latter months of 1944. He had joined Rotol in 1940, working in No 1 Assembly Shop at Main Works, transferring with the department when they moved to RD. In 1941 he joined the RAF as an airframe fitter with 11 Group Fighter Command and spent some of his time at OTU Lichfield, setting up a propeller bay to assemble hydraulic airscrews which came to the unit in boxes, unassembled. Chest problems caused him to be invalided out of the RAF in 1944, and he returned first of all to the old Crypt School in Gloucester, where he became part of a three-man team producing sectioned models of airscrews and governor units, and shortly afterwards moved once again to the 'toy factory' in Upton Street. Unfortunately, his health continued to deteriorate and he was transferred to the 'marking-off' bench, in the Experimental Shop, where he could at least sit down. After a spell in hospital, Wally returned to Experimental, where Charlie Beasant was the Shop Superintendent and remained there until Jock Maclean gave him a job in the new model shop, which was set up after the Dowty take-over in 1959. Wally, who retired in 1984, had a fund of stories from the years when he travelled all over the world, supervising the installation of the company's exhibitions.

In February 1941, the Auxiliary Drawing Office at Imperial House in Cheltenham, under Pobjoy, had been in operation for almost a year when a young draughtsman arrived from Manchester. T.D.H. (Tommy) Andrews started work at £6 per week and was soon involved in gearbox design, moving on to take charge of a small team of technicians, among them a young ex-apprentice, Roger Armstrong. When jet-propulsion became a talking point in the office, Roger recalls that Pobjoy demonstrated its principles with a blow-lamp, lashed on to a roller skate. Andrews' team had responsibility for the development of the Rotol Pressure Cabin Supercharger which was destined to be passed to Sir George Godfrey and Partners at the end of the war, when the Rotol Auxiliaries Department closed down. Norman Anderson and Vic Davis who had been closely associated with the equipment elected to go with it. Since then, the air-conditioning of high-flying aircraft has progressed to a point where it is now so commonplace that it is taken for granted by most travellers.

When Pobjoy left in 1945, as recounted in a later chapter, Andrews accompanied him, but Pobjoy's subsequent death in an air crash forced him to consider other plans which resulted in him starting his own business with E.D. (Don) Wynn, who was another member of the old auxiliaries team with similar ideas. As Wyn-struments Ltd they produced scientific instruments in premises which had been the pre-war clubhouse of the Cotswold Aero Club and subsequently the RAF Officer's Mess throughout the war. (These premises are still to be found just behind No 6 hangar on the present South Works site). George Dowty bought the business in 1954 and Tommy Andrews' subsequent career was with different companies in the Dowty Group. He was appointed Group Technical Director in the 1960's and in 1969 was offered the chairmanship of Dowty Rotol but Una, his wife, persuaded him to turn it down. He died suddenly in 1971, but his progress from a lowly draughtsman to the very top of the tree, proves the old adage, that a field marshall's baton lies in every soldier's pack.

Another draughtsman who joined the Auxiliaries Department in November 1940 was John G. Walter, who had been a regular contributor to the magazine *Punch* before the war. At the outbreak of hostilities, he had taken a nine-month draughting course in Bristol, and on completion had made his own way to Rotol (contrary to the regulations – he had been scheduled to go to Ferranti in Manchester) where he was interviewed for a job in the Auxiliaries DO by George Bubb. The office was still at Main Works at that period, but when it moved to Imperial House, John stayed behind as a kind of liaison officer. This proved quite an education in itself, as it gave him the run of the factory which he much enjoyed. His talents as a cartoonist soon became known and we are fortunate in having examples of his work in several issues of *The Airscrew*. After a couple of years he moved to Imperial House where he was generally employed on that most soul-destroying of jobs in the Drawing Office, preparing 'mods' (modifications) as they are commonly known.

In order to fly key personnel round the country, to attend various conferences and visit other companies, Rotol was allocated a General Aircraft Monospar by the Ministry of Aircraft Production, and Bryan Greensted and Ted Edwards collected the machine from Hullavington in May 1941. The Monospar was a light aircraft, powered by two Pobjoy Niagara engines (the same Pobjoy who was now responsible for Rotol Auxiliaries) and stayed with FTD until it was replaced some time later by an Oxford.

It was in 1941, that a man who was to have a great impact on the lives of his fellow workers, returned to Rotol. Frank Humphries started in 1938, as a fitter in the propeller assembly shop under Harold James and eventually went on to complete almost 40 years service with the company. As a reservist he was called up in 1939, serving with the Tank Corps until being invalided out in 1941. On his return he found that Rotol had been completely transformed. New buildings had gone up and the whole atmosphere had changed dramatically. A great sense of urgency and purpose now seemed to prevail throughout both the works and offices. Frank was not alone in noting this, as almost everyone employed at Rotol during the war made substantially the same comment. The man largely

ROTOL·AIRSCREW·SIDELIGHTS, No 5.

THE FIRST SUCCESSFUL (IF SOMEWHAT HAZARDOUS) ATTEMPT TO PROVIDE PROPELLERS WITH A

VARIABLE·PITCH.

responsible for this transformation, in many people's opinion, was R.H. Coverley.

Frank Humphries was a well-known figure at Rotol, to both management and workers alike. He was a keen trade unionist, serving on almost every committee of the Amalgamated Engineering Union (AEU) both at district and branch level, and it was natural that he should take on the office of works convenor at Rotol. Although Rotol had a remarkably strike-free record, Frank saw to it, that wherever possible, the company were made aware of their shortcomings. This manifested itself in small but significant ways. One of his members collapsed and died at work and a collection was made among his workmates for his widow and young children. It only amounted to a few pounds, but Frank realised it was all she was going to get. From that moment he resolved to put his efforts into persuading the company to provide a pension fund for manual workers, as well as staff. He makes no claim to being the key figure in the eventual realisation of that dream of nearly fifty years ago, but today there is no distinction between works and staff personnel in the Pension Scheme operated by the Dowty Group. The Rotol Pension Fund was not set-up until the war was over, although it was the subject of some discussion by the Directors in June 1945, when Sidgreaves suggested that it should be, 'held over until the staff was in a more stable condition'.

On another occasion, Frank was detailed by Eric Searles, who was then Superintendent in the Gear Shop, to accompany him to the funeral of one of the workers who had passed away. When they arrived at the graveside, they were saddened to find they were the only two mourners. Eric reported this back to the company and from then on it became official practice for the company to send a representative to all employees funerals. Frank also commented on the enlightened attitude that Rotol adopted towards disabled and blind workers, as both of these categories were put on the pay-roll at an early stage of the company's development.

According to Frank, the practice of working on Good Friday and having the following Tuesday in lieu, began during the war. This was to facilitate travelling, as a large number of employees lived quite a distance from Gloucester and the arrangement meant that trains were not so crowded. The agreement had already found favour with Gloster Aircraft Company and some other firms in the area, before being adopted at Rotol. Bank Holidays and the annual one week's holiday in August, were unpaid for works personnel, and were always a bone of contention between the manual workers and office staff. It is only in more recent times that this difference has been largely eradicated.

During the research for this book, the author spoke to many ex-employees who recalled Frank Humphries. Without exception, he was remembered as a man of his word, always ready to defend his members and improve their lot whenever he could. He fought like a tiger if he thought an injustice needed righting but had no time for any frivolous complaints. As a trade unionist, and a staunch protector of his members' rights, he had few equals.

The Welfare and Social Club, as it was known originally, was first mentioned at the February 1938 meeting of the Directors, when they agreed it would be a desirable feature of the new company and an attraction to new employees. It was modelled on the lines of the Rolls-Royce Club and at the April 1938 meeting, a grant of £550 was made to enable the scheme to be inaugurated.

The first Rotol Club House at Main Works was situated in what is now the Cheltenham Car Park and had originally been used by the Cotswold Aero Club. It was a popular and convenient meeting place, rather too popular if my informants are correct, as it was sometimes necessary to drag reluctant night shift members from its seductive charms. It was burnt down in the early war years and, according to eye-witnesses, the fire quickly got a hold, completely destroying the wooden structure in a short space of time, watched all the while by an interested work-force, who had just completed the day shift. Another wooden building on the Sports Field (now used by the Bowls Club) came into being, but that too was extensively damaged by fire, sometime in the 'sixties.

This is perhaps an appropriate moment to introduce the reader to the Rotol Welfare Department. Under the direction of E.O. Cameron the Company Secretary, the Welfare Office was an extremely busy place during the war years. It was originally located in some huts adjacent to the Hatherley Lane gate, where the present Laboratory now stands. H.A. Salkeld was Chief Personnel Manager, C.N. Taylor the Chief Industrial Welfare Officer, J.W. Lowley the Welfare Supervisor and Miss Evans and later Miss Ashworth as the Women's Welfare Supervisor. The Society initiated a scheme at an early date by which an employee would receive benefit by paying a small voluntary sum each week. The original scale of benefits operated for some time but changes were introduced in November 1942, when the deduction became 10d. (4p) per week (juniors and apprentices paid 5d.) from employees wages, with a contribution from the company of 3d. The scheme was initiated to decrease hardship and provide a small income of £1 per week in times of sickness and there were other benefits, including the use of the Rolls-Royce and BAC convalescent homes by Rotol (and at a later date, British Messier) employees, should they be needed. The Welfare Society continued to operate this voluntary scheme until the company became part of the Dowty Group.

The Welfare Office encompassed a great deal of what is now handled by the Sports and Social Office, including the sports facilities and Club House. In addition to looking after the leisure hours of the employees, it was also responsible for all matters concerning the employees welfare, including advice on Income Tax, supplied by Mrs Anne Webster. She was appointed to this post in November 1942, and in a little over twelve months answered more than 4,000 queries. F.C. Bunn had the task of organising daily entertainment in the canteen during the lunch break, not all of it being fully appreciated by the audience according to some of my informants, although the boxing matches always seemed to get a good attendance.

In the early part of 1938 the first temporary canteen was located in a wooden hut not far from the Hatherley Lane gate, which was for men only, the ladies having a separate canteen, although why this should be so is not clear. It catered for Executives, Senior Staff, Staff and Works personnel and the same hut also housed the Labour Office and Police Lodge. The need for a more permanent canteen was obvious and a tender from W.J.B. Halls of Gloucester for £5,100 was accepted on 19th May 1938, with an anticipated completion in twelve weeks time. The canteen was considerably enlarged at a later date, possibly soon after the war had begun, into the form which most people will remember. There was no segregation of the sexes, but works personnel had a separate section from the staff, although they all shared a common menu. The executive staff had their own mess, where the company provided free lunches and drinks for several years.

In the works canteen was a stage at one side, which was enlarged in 1943, reputed to be 'capable of holding a total of 20 performers in comfort' and when one of the ENSA parties was not available, employees would sometimes provide their own entertainment. Jack Sullivan praised the singing of Herbie Rickards, who often sang 'Jealousy', and the rendition of 'Gold mine in the sky' by Jack Williams always provoked some ribald comment from the audience. The piano playing of Jack Gronow from the Crack Test Department was also much appreciated. To complete the quartet, Jack Shaw sang the popular ballad 'On the road to Mandalay' on numerous occasions, this being the only song he knew – or so his mates claimed.

Food rationing was one of the most lasting impressions of wartime life in Britain and all over the country a barter system sprang up whereby someone who didn't use sugar would trade it with a family that had excess tea, or any of the other foodstuffs that were surplus to requirements. Rationing also meant that eating at home had to be carefully eked out, and the canteen service, providing daily meals at small cost, was used extensively during the war. The main meal cost 8d., pudding or a sweet was 3d., and cup of tea was 1d. bringing the total cost to 1/- (5p) and there were reduced prices for juniors. In order to purchase a meal, one had to queue at the kiosk and obtain a ticket before queueing once more at the counter where the food was served. Everyone provided their own knife, fork and spoon, with a mug or cup for tea, and washing-up facilities were available. (This extended into quite recent times and was only done away with when the new canteen was built at the end of the 1970's.) In the early part of the war, two of the canteen staff came round to all the machinists on the Night Shift to get their orders, but this may only have been a temporary arrangement.

The canteen was originally operated as a commercial concern by a company called Barker, but in the middle of 1942 the Directors agreed to a plan put forward by the Shop Stewards, where the canteen was run by a sub-committee composed of staff and works employees, with a chairman appointed by the management. In 1943, the chairman was C.H. Davis, the Labour Manager, and the thankless task of secretary was taken on by J.H. Lapham. Mrs Bragg had been the original supervisor in 1938, but left to join NAAFI eventually returning to the company to become Canteen Manageress. The canteen run by a committee seems to have worked very well throughout the rest of the war, and even made a modest profit each year. At the end of 1945, fewer employees meant that it began to operate at a loss and this may have been the reason for the scheme ending soon after.

Charles Baker was the Welfare Secretary and in 1941 he also took on the post of Personnel Manager for a while. A copy of *The Airscrew* at the time, offered congratulations to him on his new appointment and went on to say, 'We do not always see eye to eye with him and do not expect to (surely this has been the lot of every personnel manager since time began – author) but we have come to realise that everything he does is with the object of assisting the employees.' When he left the company in July 1944, he was presented with a tankard and salver, at the Rotol Cheltenham Social Club situated in premises off the High Street, thought to be in a passage running alongside where the present Marks and Spencer's Store is now situated. The club had been opened by Cameron on June 13th 1944, as a social centre for the Rotol employees living in the area but did not continue for long, due in part to the end of the war in 1945 and the general financial cut-back which followed.

The Medical Services at Rotol also came under the jurisdiction of the Welfare Office, and here we meet a lady who will be well-known to all Rotol workers. The nursing career of Sister V. Rees commenced at the Great Western Railway Hospital at Swindon, followed by further all-round training at Cheltenham General Hospital. She was made a member of the Royal College of Nursing in 1940 and probably came to Rotol soon afterwards, where she was to remain throughout the rest of her working life. In wartime, she had seven or eight nurses in her charge and the Medical Room in which they spent their day – and night – has not altered much since it was first built in May 1941. The nurses have always had a busy life at Rotol dealing with a variety of cases from minor injuries to much more serious problems. At one time it was noticeable that the Medical Room was always more crowded on a Monday morning when apprentices and those who had been engaged in boisterous activities during the weekend came to have their aches and sprains administered to.

There were some significant executive staff changes during the latter part of 1941, the Production Manager B.W. Angel, the Chief Engineer H.N. Charles and Gowers the Chief Buyer, were released from the company's service. There must have been a very good reason for them leaving as it was by no means a simple matter to leave one's employer during the war in the normal course of events. In their place came two promotions from within the organisation, C.J. Luby to Works Manager and L.G. Fairhurst became Chief Engineer. E. Danvers was appointed Chief Designer, replacing LGF and the position of Chief Buyer was taken by S. McKiernan.

Mention has already been made of the Project Design Office, situated over the Spinner Factory in the High Street, where the young designers were working on future Rotol products that would probably not be seen in the assembly shops for some time. B.V. (Vic) Holberton, at the age of 23, must have been one of the youngest. He had originally worked at Hoffman Bearing Co at Chelmsford and more recently with Aircraft Components at Arle Court. Vic was among the first Project Designers to be so-called by Rotol when he joined the team led by Tom Godden in 1941. One of the earliest designs with which he was associated, was an idea for changing the method of altering the pitch of the propeller blade from a moving cylinder and stationary piston, to a moving piston with a fixed cylinder. This became the standard for all future Rotol propellers and was eventually patented, carrying the young Holberton's name – after some debate – as it was not then company practice to have a junior employee's name on a patent. He was also concerned with a modification to the pressure relief valve in the governor unit, increasing the pressure on the coarse pitch side of the propeller operating piston, thus preventing the propeller from driving the engine, when the aircraft was in a high-speed dive. This had been suspected of causing a number of crashes prior to the introduction of the improved design. Rotol bought the patent rights from Holberton and Bishop (his co-designer) for £75, which they divided between them. At a later date, due perhaps to the frustrations and restrictions and the long working hours of wartime, some disagreement ensued between the design staff, resulting in several changes being made. A few members of the team were seconded to the offices at Main Works but Holberton remained in the Project Design Office.

A vitally important part of the airscrew has been ignored up to this point, the airscrew hub, housing the pitch-changing mechanism and into which the blades were located. It arrived in the factory as a solid steel forging, weighing, for a three-bladed airscrew, something in the region of 350 to 400lb but after final machining it was reduced to between 40 and 60lb. This was both time consuming and wasteful in material and an alternative solution was urgently sought. The first successful idea was to form the hub in two halves, using steel forgings, machining and subsequently bolting the two halves together. By the middle of 1941, a second process had been developed by the company, involving pressing the hub in two halves from one inch thick plate and then flash-butt welding the two parts about the centre-line. Many thousands of hubs were produced in this manner, conserving approximately 150lb of steel per airscrew at a saving of £17, and the technique was adopted by other manufacturers.

In order to check the suitability of hub design, heavy weights would be fixed into the hub arms, mounted on a whirling rig in the Test Hangar and then spun to a high rpm On more than one occasion these weights separated from the hub, flew through the roof, and eventually landed in various parts of the factory, fortunately without causing any injuries.

Ideas for speeding up production were always under consideration and often resulted in saving time and material, but not all were completely successful. On one memorable occasion, instead of the more usual forging, an experimental hub was made from a steel casting, to find out how this might stand up to stress and fatigue. It went right through the complete machining process, but somewhere along the line the paper-work was lost. As it was a single item in a large batch of forged components, identical in every respect except for the material, it was impossible to distinguish the 'rogue' hub from the rest. There was no question of scrapping all the good hubs, so the company had no option but to release the batch. Somewhere, a Halifax bomber flew with a cast hub, but no one knew which it was and nothing more was heard of it, but LGF had the unenviable task of admitting the company's error to the MAP at the next meeting. The idea of cast hubs was not proceeded with.

George Pettitt (Clock No 612) worked on the Hub Line and recalled some memories from the early days. The works buses from Gloucester dropped their passengers off in Hatherley Lane, and they would then run across the field where Heat Treatment and No 2 Shop now stand, avoiding the numerous cow 'pats' if possible (the local farmer still used the field for grazing his cattle at this time) to reach their clocking station. George worked on the grinding section, which had an issue of free milk and extra clothing coupons for their overalls. The coupons were always in great demand from the ladies, who could never get enough of these most treasured items. In common with others, George's strongest memories of the war years are of a comradeship among the men and women on the factory floor which helped them to keep going through the long hours.

Towards the end of 1941, Fedden once more opened a discussion on what he considered to be a low level of airscrew production throughout the country, currently running at some 27,000 per annum. He suggested that Rotol and De Havilland should make a joint approach to the Ministry, pointing out their concern over the inablility to meet projected target figures and the necessity of increasing production by whatever means were available. Rotol had been placed in some difficulty by the MAP airscrew programme continually being updated in each of the latter months of 1941 and despite efforts to increase output by additional labour and new buildings, the current programme looked like being short by 750 per month. These figures were looking ahead to mid-1943, in the meantime Coverley reported a general re-distribution of labour in order to convert the large stocks now held into completed airscrews and concentrating on parts which were still lacking. It was agreed that Hives, Fedden and Coverley would meet Sir Charles Craven, an MAP official, during the following week, to point out the problems being experienced by Rotol and insist that another factory would be needed to meet the programme. These discussions must have been fruitful as by the date of the next Board Meeting a formula for making certain of increased production had been thrashed out. These measures included manufacture of airscrew parts in America, production assistance by No 1 Shadow Group and a new factory under Rotol control. This latter enterprise was to be situated in Tottenham, where a factory had been requisitioned and £1,750,000 sanctioned for the necessary machine tools and equipment, all financed by the Ministry.

Over the years, Rotol has been visited by many important people, and occasionally by VERY IMPORTANT PEOPLE. The company is fortunate in still possessing the visitors book which was kept by the commissionaire in the reception area of the 'White Block'. By dint of much detective work and cross referring to the copies of *The Airscrew*, it is possible to decipher some of the signatures, although it has to be admitted that the majority have so far defied identification. The following tale may come into this category; Eileen Hunt (née Alder), worked in the typing pool situated in the 'White Block' and was sometimes privileged to enter through the front door – when the commissionaire was not looking. She was in the company of her friend Iris Brown one day, when they came past the commissionaires desk, which was unguarded. On the top lay the visitors book, invitingly open. 'I dare you to sign it', said Iris and – needing no further encouragement – Eileen promptly did so. Sad to relate, despite making a careful search through the book, I have not been able to find her name on any of the pages – maybe she wisely decided to sign someone else's name instead!

The book does not commence until February 6th 1941, which is a pity as many interesting people must have visited Rotol before that date and others did not sign the book for a variety of reasons – perhaps it was mislaid at some period. One of the easiest signatures to decipher is that of HM Queen Mary, Gloucestershire's most famous evacuee, who – together with a reputed 55 servants – stayed at Badminton House, the country home of the Beaufort's and was often seen in the surrounding district. She visited the Rotol Repair Department in Gloucester on December 9th 1941, but apart from a number of photographs in *The Airscrew* very little is known of her tour on this occasion. Two years later when she came to the Main Works, the magazine produced more photographs and a full report of her visit will be related in the chapter dealing with that period. Queen Mary was not the first Royal visitor however. Her son the Duke of Kent, a Group Captain in the RAF came to Rotol early in 1941, although the exact date is unknown. We know even less of his visit, except that he toured the factory and had his photograph taken with Coverley.

Shortages of essential supplies began to occur more frequently as the war progressed and high on this list were ball bearings, giving rise to a certain amount of anxiety on the part of the Rotol management. In April 1942, consideration was even given to the setting up of a brand new bearing company, in order to overcome the shortage. Correspondence with the Minister of Aircraft Production failed to resolve the problem, which threatened to assume massive proportions by the end of 1942, by which time it was forecast that 2,000 airscrews would be held up for lack of suitable bearings.

As a first step, Rotol Repair Department set up a reclamation scheme at ARD in Gloucester in the summer of 1942, with a special department equipped with the latest type of machinery to enable the reconditioning of all forms of bearings to be undertaken. At the time Rotol was believed to be the only organisation outside the bearing industry to possess this facility. By the end of the war, it was calculated that over 78,000 stacks of bearings had been reconditioned by the ARD. As an illustration of the large number of bearings used, each blade of the three-bladed airscrew had a top, centre and bottom bearing track in one stack, weighing about 11lb with a diameter of almost 8″ at the blade root.

Gearbox production continued at Rotol under the dynamic personality of Eric Searles, who was noted for going everywhere in a tremendous hurry. As John Chilman is reputed to have remarked one day on the telephone, 'Hang on a minute, I've got a fly in here, going round like a blue-arsed Searles.'

Dick Smith, a draughtsman in the Main DO reminded the author of one of Eric's habits, when he would stride into the office and proclaim in a great voice, which carried to everyone within earshot, terrifying all and sundry (but YOU in particular), 'When am I going to get an answer on this query – I've got ten machines standing idle, waiting for you.' The fact that the problem, whatever it might be, had only been handed to you five minutes ago, was no excuse and most of the young draughtsmen – including myself, were completely unnerved by Eric's display.

This particular approach didn't work with everyone. Pete Walby was only sixteen in 1942 when he started work in the gearbox shop, and readily admits he was somewhat short tempered. One day, Eric, in his forthright way, asked Pete, 'What are these bloody casings doing on this bench', pointing to the offending articles. The reply that Pete gave was somewhat short, sharp, and to the point, and Eric Searles hauled him off to the Personnel Department where he was severely admonished by Harry Davis, the Manager, for using bad language.

Following this incident, Pete was sent to the Spinner Factory in Cheltenham, where another row with the supervisor ensued, over some trivial matter. This time, Harry Davis was more severe, insisting that Pete should somehow control these outbursts in future, or face the inevitable consequences. Pete was offered and gratefully accepted a job in the Test Hangars back at Main Works, working on the night shift. He seems to have settled down in this department and enjoyed

The Tool Room in 1939, part of No 1 Machine Shop.

Rotol's first cricket team in 1938. L.G. Fairhurst third from right in rear row. Seated right is W. Tattersall and next to him is R. Parsons.

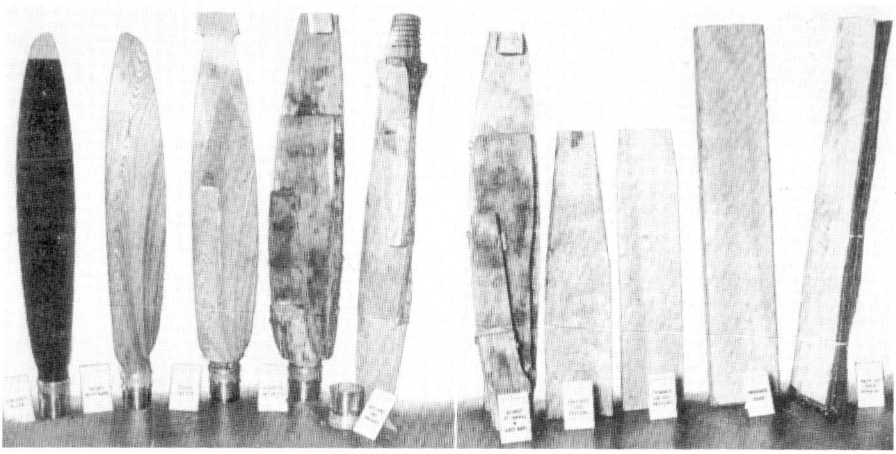

Twelve stages in the manufacture of a Hydulignum wood blade. *Right to left*: pack of veneers, consolidated board, board trimmed for side pressing, after side pressing, assembly into glued block, block butt-ended and threaded, adaptor mounted, blade rough shaped, finish white-shaped, finished blade.

Copy-milling a wood blade (top) from a master blade (bottom).

Wartime airscrew assembly shop.

Balancing a wood blade airscrew. Shallow pits allow the blades to rotate and bring the hub to working level.

Building the new 'time recording station' – later the gatehouse. The RAF hangar in the background, on the other side of the Gloucester Road, became Rotol No 6 Test Hangar in later years.

The 'White Block', well camouflaged in 1942.

Putting in the foundations for the original technical block. Note the air raid shelters on the right.

Building the technical block. The petrol pump in the foreground was part of a garage on the corner of Hatherley Lane.

Mrs Emma Lewis (née Day) driving a Douglas truck, with her cousin Florrie Evans sitting behind her. The camouflage netting which covered large parts of the factory during the war can be seen overhead.

The Rotol Canteen Committee, 1943. *Left to right*: Messrs Arnold, McCullum, Mrs Taynton, Messrs Lapham, Davis, Parker, Mrs Greenland, Messrs Taylor and Pritchard. (Mr Wright – absent.)

Members of the Rotol Amateur Dramatic Society in their production of 'The Roof', given in Feb. 1943 at the Cheltenham Opera House.

The Repair Works Social Club organised their first dance on June 8th 1942. Bill Warner stands third from right.

Queen Mary paid a visit to the Repair Works in Gloucester on Dec. 9th 1941. In the lower left photograph she is accompanied by Stammers and Bleaney, on her right.

R.H. Coverley and the Duke of Kent in February 1941, walking near the Hatherley Brook.

R.H. Coverley with Prince Bernhard of the Netherlands and LGF and Stammers on the right, seem to find something of great interest just off the picture. The location is thought to be the Repair Works in Gloucester.

'Glensanda' – offices for the Publicity Department.

Imperial House – the Auxiliary Design Offices near the Gordon Lamp, Cheltenham.

Battledown Manor – guest house in Charlton Kings.

Three of the properties used by Rotol during the war, photographed in 1990.

his stay, except for the excitement when a propeller shed a blade and it flew through the roof. ('Burst' hubs was not an isolated occurrence, according to my informants). Pete joined the Grenadier Guards in 1944 and did not return to Rotol until 1952, when he resumed his career in the Test Hangar. Life in the army had changed him in many ways and he no longer lived quite so close to the shadow of the sack. At any rate, he went on to serve the company until early retirement in 1983.

In the middle of 1942, another round of staff organisation changes were made: T.W. Morgan became Production Controller, A. Jackson the Works Manager and C.J. Luby was appointed to the position of Technical Production Engineer. At the time, Luby was in the process of resigning his position at Rotol but Fedden advised him to retain his connection with one of the parent companies, if he decided to leave. He must have considered the matter further, as he was to remain with Rotol for some time. However, Jackson left after a few months, (his post was filled by Les Law before the end of the year) to take an appointment with Gloster Aircraft Company, and his secretary, Mrs Eileen Robjohn (née Walker) went with him. She remained with Jackson for a time, followed by a spell with the Air Regulation Board in Cheltenham but was persuaded to rejoin Rotol after receiving a plea from Miss Ashworth, who was having some difficulty in finding a suitable replacement secretary for a 'fiery little man', named Pobjoy. Eileen was to be associated with Rotol and later British Messier for many years, acting as secretary for many of the top technical managers and will be especially remembered as one of the country's foremost exponents of ballroom dancing. At one period during the war the Welfare Society organised a regular dancing session during the lunch break, which was welcomed particularly by the ladies, at which Eileen assisted.

Sir Roy Fedden made his last attendance as a Rotol Director at a Board Meeting on September 10th 1942. Fedden, who had received his knighthood in the 1942 New Year's Honours List, became Technical Adviser to the Minister of Aircraft Production until 1945, leading missions to America, Italy and Germany and among his last duties was that of consultant to the Dowty Group from 1953 to 1960. His place on the Rotol Board was taken by N. Rowbotham.

When Coverley came to Rotol in 1939, he occupied a suite of rooms at the Royal Hop Pole Hotel in Tewkesbury – where, according to his secretary – they served home-made apple pie, which he could not resist. He remained at the hotel for a number of years, until he persuaded the company to lease Battledown Manor in Charlton Kings, near Cheltenham, where their important visitors could be entertained, making them independent from the local hotels. Coverley had first put the proposition at a meeting of the Board in May 1940 but on that occasion it was turned down. In January 1942 he brought the subject up once more and this time it was accepted. Coverley moved in shortly afterwards and stayed until his resignation in 1945. The Manor was situated in a commanding position, with extensive grounds overlooking one of the finest views in the area. During the war years it was the setting for many memorable parties and celebrations, where the staff were reputed to have worked wonders with the meagre food rationing regulations.

One other resident at Battledown Manor, at least for a while, was Maria Luth (or Mrs Stormonth as she was known at Rotol). She had been transferred from

London towards the end of 1942, to become the MAP Programme and Liaison Officer and as a gesture to the Ministry, Rotol offered her a room at Battledown Manor, which she accepted.

By all accounts Maria was a very charming young lady. Her poise and elegance caused many a male heart to flutter, and several of my informants have commented on her dazzling appearance at the Rotol Ball and other social occasions. She had originally trained as a ballet dancer, and was one of the first dancers to perform on television and had also appeared in the original production of Noel Coward's *Cavalcade* at Drury Lane. Altogether, one can appreciate that Mrs Stormonth was a welcome celebrity in the wartime atmosphere at Rotol and with her background it is not surprising that she should be in great demand when concert parties were staged at Rotol. Maria is first mentioned in *The Airscrew* on July 8th 1943, when she sang on the canteen stage and again at the end of 1943, as producer of a show called the 'Rotol Christmas Caravan', again in the canteen. She may have made other appearances, but these are the only ones to be recorded by *The Airscrew*. Maria eventually moved to the Hordern-Richmond factory, where she doubtless cast her spell over their workforce in much the same fashion as she had at Rotol.

Vernon Christie joined Rotol in May 1939 and was Section Leader in the Project Design Office. He and his wife were living in Greenway Cottage, situated almost opposite the drive leading to Battledown Manor, where they kept some goats on the back lawn. While working in the office one morning, Vernon was somewhat surprised to receive a telephone call from RHC asking if he and his wife would come to dinner at the manor that evening. The invitation was duly accepted and Vernon and his wife, somewhat apprehensively, turned up at the appointed hour and were warmly wlecomed. They dined in a fashion that was a decided improvement on the normal wartime fare and at the end of the meal, over coffee, RHC casually asked if they would consider taking a paying guest, namely Mrs Stormonth. It appeared that it was considered a somewhat embarrassing situation having Maria and RHC living under the same roof, so a solution had to be found. Although the Christies were not too keen on the idea since their first baby was only a few weeks old, after some hesitation – but perhaps not too much as RHC was the Rotol Controller – it was agreed. Maria moved in and stayed some two years, proving to be a very kind and considerate guest. Travelling to London almost every weekend, she often returned with very acceptable items of food which were unobtainable in the local area.

As a postscript to this story, when the Christies were on their way to the dinner engagement, Vernon noticed the lawn of long, lush grass. Having yielded under pressure over Mrs Stormonth he rather diffidently broached the subject of pasturing his goats on the Manor lawns. To his great surprise RHC agreed and from then on the goats were taken up to the Manor every morning and brought home again each night.

Vernon was a member of the Rotol LDV and contributed the following story:

> Our group were issued with two vehicles produced by the Standard Motor Company, called Beaverettes, for defence of the factory. These were essentially a basic Standard Eight motor car, fitted with a Bren gun and armour plate. Someone

decided that a further two were required and volunteer drivers were called for, to go and collect them from a Maintenance Unit near Manchester.

Jim Boyd, Peter Arrowsmith, Derek Waters and myself were chosen and set off by rail, to be met at the depot by an Army officer, who asked if we would like a practice drive around the camp before taking them on the road. The offer was politely turned down, with the explanation that as we already had a couple on strength, the four drivers were quite confident of driving them back to Rotol.

We quickly changed our tune when confronted with, instead of the little Beaverette, great armour-plated monsters, each weighing about ten tons and fitted with a 20 mm Hispano gun, plus a great quantity of live ammunition. Swallowing our pride, we agreed to a practice drive, mainly to familiarise ourselves with the 'crash' gearbox and after a time, began our return journey.

By this time it was well into the afternoon and obviously Rotol would not be reached before night fell. For some reason we were forbidden to travel at night, so one of the party telephoned the factory LDV commander for instructions, who told us to find an army camp and lay up there until morning. With some difficulty, we located a tented camp near Rugeley and asked the officer of the guard if he could give us shelter for the night. Initially he was somewhat reluctant, as his outfit had nothing so grand as these vehicles being driven by a group of 'civvies', despite the fact we were in LDV uniform. However, he soon relented and we spent the night under canvas, setting off for Rotol the next day. Unfortunately, the two armoured beasts proved completely useless for factory defence, and were eventually exchanged for something more suited to the purpose.

The Airscrew magazine of November 1942 carries some information concerning the apprentices of the time, and these are further amplified in the issue of March 1943, which are of special interest because we first meet some of the lads who were to stay with Rotol through all of their working lives and whose names will be familiar to many. The first Apprentice Prize-Giving Ceremony took place in the Staff Canteen on February 18th 1943, when twelve 'embryonic engineers' received engineering books and drawing instruments given by the management. Coverley had been detained in London and sent his apologies, in his stead Stammers presented the prizes. The main recipient was Roger Armstrong, who has already been introduced to the reader in an earlier chapter, and it fell to Roger to propose the vote of thanks on this first occasion, after which the assembled company adjourned for tea. Among the other prizewinners was Bill Edden who received a National Certificate, with six 'firsts' and one 'second' Class pass in the London City and Guilds Machine Shop Engineering Examination and in addition won the 'Dowty' prize on two other occasions. Eddie Cowley completed his apprenticeship in 1943 and was the first boy to be awarded a practical prize because of his ability in that field – he was presented with a very useful micrometer. He went on to work in the Experimental Shop and then the Jig and Tool DO under Bill Jones and recalled that he took most of his annual holiday in half-days in order to play football for Gloucester City and Cheltenham Town, before becoming player/manager of Cinderford Town. Another 'Dowty' prize winner and recipient of a National Certificate with distinction in Engineering Science was P.R. (Phil) Smith, who had started at Rotol as an office boy in the sub-contract department in October 1939. After completing his apprenticeship, Phil worked in the Production Department Jig and Tool DO and later in Process

Planning before transferring to the Works Engineers Office, where he spent the remainder of his working life.

There was never a shortage of boys eager to take up an apprenticeship, despite the low wages of the time. John Hunt started in March 1942, when the apprentice rate at age 15 was still only 15/- (75p) per week. John spent three months in the Apprentice School, followed by nine months in No 1 Machine Shop on high-speed lathes and then three years in the Gauge and Toolroom, completing his apprenticeship in 1946 as a skilled toolmaker. He was anxious to become a draughtsman and managed to arrange a transfer to the Modification Section of the Main DO on a three month trial basis, before graduating to the Hub Section under Fred Ault. When the Fuel Flow Proportioner Section was started in the early 1950's John was chosen as Section Leader, at the early age of 25, and went on to complete over forty years with the company in various responsible positions.

The scope of the Rotol Apprentice Scheme was increased as time went by, with the addition of a Junior Technical School at Main Works which was opened on September 21st 1943. It had the object of raising the standard of apprentices between the ages of 14 and 16 years who had joined the company from Elementary Schools, enabling them to start their National Certificate Course at the local Technical Colleges with a solid grounding. The School made full use of the talents of several staff personnel, who served as part-time lecturers, all of whom were reputed to be volunteers. The majority came from the Research Department and included Don Bishop, Alf Mapp, Norman Ramsey, W. Gibson, M. Rees, E. Morgan and Ken Boydell. Other offices within the factory also supplied A. Barlow, Jig and Tool DO, L.B. Jones from the Stress Office and S.W. Beard of the Main DO.

A Rotol apprenticeship, although desirable, was not the only admission to the company. Boys of fourteen, straight from school, began work in the Welfare Department as messengers or office boys, delivering mail and other articles throughout the factory. George Kent quickly learned the whereabouts of all the departments throughout the site, when he started in November 1942. His duties included frequent trips to the canteen, to get tea and buns for the office, washing up the cups and saucers, as well as collecting and delivering mail. Covering such a large site, he estimated he walked some six to eight miles a day. One of the most poignant recollections of his time as a messenger boy was being permanently short of cash, as he gave his mother 17/6 (87p) out of his weekly wage of just £1.

He lived at Withybridge Road and every day he walked to and from the factory. This meant that he came past 'The House in the Tree', a well known hostelry, where there was a gun site manned by soldiers and ATS girls. He always remembered one particular Christmas, when an officer approached him and politely asked him to stop kissing the girls every time he went by!

It was appreciated at an early stage that Rotol alone would not be able to produce all the wood blades needed for the airscrew programme and in order to ensure sufficient quantities were available factories in the North of England were taken over and equipped with the necessary machines. The workforce was either recruited locally, or directed there by the Ministry of Labour, but a number had to be sent from Main Works to teach the new recruits the unfamiliar techniques. The first to commence production was near Trafford Park, Manchester, originally producing the complete wooden blade, but settled eventually for supplying only the glued blocks from which the blades were finally machined. By 1941, the combined capacity of the Trafford Park factory and another at Stockton was 640 Jablo blocks per week. Schwarz blades were also being made in Canada, where an order for 2,000 was placed in the same year, but as one of the first operations in the manufacture of wood blades was fitting the steel adaptor to the root end, before rough profiling the block, Rotol were asked to urgently provide 1,000 adaptors in order to get production moving.

Another factory was taken over in 1942 at Ossett in Yorkshire, previously known as 'The Mill' because it supplied 'shoddy' or re-cycled wool extracted from rags for the woollen industry. At Ossett the blocks were turned into the finished blade, to be sent back to Main Works for assembly in the completed airscrew. The whole process, from glueing of the blocks, up to the finished stage – which included the covering process – took some six to eight weeks, allowing adequate time for stabilising between the various stages of manufacture.

The Blade Production Manager, responsible for both the Trafford Park and Ossett factories, was Harold James who came to Rotol from BAC and as already mentioned, was one of the first employees at High Orchard Street in 1937. He had done the actual hand-shaping and machining of the metal blades used on the long-range Wellesley, being promoted to Assembly Shop Foreman at Main Works in 1938. A.E. Marsh was the Shop Superintendent and both he and James had first moved from Main Works to Trafford Park, the first Rotol wood blade factory, in 1940.

Another was Don Rendall, who had started in 1938 as one of the skilled carpenters in the Experimental Shop at Main Works, who numbered around 24 on the strength. In those early days, shaping the wood blades was carried out entirely by hand, using the traditional tools. Don Rendall also first moved to the Manchester factory in 1940 but transferred to Ossett where he became the Shop Foreman. When he was asked in 1989 for his memories of the years he spent with Rotol Airscrews, he could still recall drawing number RA.4067 as being the wood blade for the Hawker Hurricane.

From the moment the Ossett factory began, it was in the forefront of wood blade manufacture. Just how many airscrew blades came from their workshops is not known but it must have run into many thousands. According to the records, Rotol built over 30,000 airscrews for the various marks of Hurricane and Spitfire alone (the Mk V and Mk IX Spitfires accounted for more than a third of this total) and each one had three, four or five blades. *The Airscrew* reported that the factory employed between 75 and 80 inspectors, 35 of them women, and it seems to have been a happy community as can be seen from the articles contained in the pages of the magazine. During 'Salute the Soldier Week' at Ossett in June 1944, the little company received visits from civic dignitaries and celebrities among whom was Jean Batten the famous airwoman, who toured the factory and delivered a short address to the workforce. Ossett raised over £3,624, beating their previous endeavours in 'Wings for Victory Week' of 1943 by over £1,000. During the lunch break, some high prices were paid at a mock auction, as recorded in *The Airscrew*, – shirts at £19, sheets at £100 and Gloucester eggs at £10 per dozen, bringing in a total of £1,017.

Electric Propellers and Further Changes

Prior to the outset of the Second World War, Rotol had made a few electric airscrews under licence from the propeller division of Curtiss-Wright Corporation in America. The original licence belonged to the Bristol Aeroplane Company and was transferred to Rotol in 1937 at a cost of £10,000. The engine companies were always somewhat reluctant to allow the airscrew manufacturers to interfere with the engine lubricating oil system and the fact that the electric airscrew had no need for access to the engine oil was one of its attractions in their eyes. However, some confusion surrounded the start of the electric propeller programme at Rotol. At the Directors meeting on 19th May 1938, Fedden stated that BAC 'were expecting to fit electrically operated airscrews to their new types of engines,' whereas at a later meeting Hives said he would not recommend electric airscrews on Rolls-Royce engines. Obviously the Directors were not unanimous at this juncture but Fedden maintained that the Rotol version was an improved design, and in the fullness of time his faith was justified, as a total of just over 7,000 Rotol electric airscrews were eventually fitted to the many versions of the Wellington bomber, powered by Bristol Hercules air-cooled radial engines.

To give some indication of the uncertainties that faced the company in the early months of its life, the Directors received a report in July 1938, probably prepared by Stammers the General Manager, that 'Rotol is proceeding with sixteen electrically operated airscrews, four of which were ordered originally from Bristol by the Air Ministry and eight had been ordered by aircraft constructors since that date.' Some doubt was expressed as to the wisdom of accepting orders from the AM 'as we were contracting to sell a product which had not been tested in production and about which we knew very little,' according to the minutes of the meting. In September, the Board stopped work on all but the four ordered by the Air Ministry and gave Stammers a reprimand for taking on the others without proper authority.

A year later, in September 1939, Fedden returned to his championing of the electric airscrew by stating that 'considerable orders might be forthcoming for the Hercules engine if the Air Ministry and Aircraft Constructors could be persuaded to accept Rotol airscrews – particularly in the case of the Short Stirling which had been designed for Rotol airscrews but had to be changed to the Hamilton type, owing to Rotol not proceeding with the Hercules size electric airscrew for various prototypes, including the Stirling.' An electric airscrew satisfactorily completed a 250 hour endurance test in September 1940, but by this date the Air Ministry had decided that the Stirling and Halifax bombers, both powered with Bristol Hercules engines, would have De Havilland Hydromatic propellers.

Pitch-change in the hub of the electric airscrew was effected by an electric motor, part and parcel of the original Curtiss design, and should have been capable of being used without modification. Unfortunately, due to the dimensions and tolerances on the drawings, or rather lack of them, each motor had to be hand-built, and parts were not interchangeable. Readers familiar with this state of affairs will readily appreciate that this was unacceptable on mass production and Rotol had to immediately re-design the motor.

At the end of 1941 the new drawings for the electric motor were given to Blackwood, Morton and Sons at Kilmarnock, which – like Rotol in 1937 – had also been a carpet factory but now converted to a Rotol sub-contract depot for electric airscrew parts. As an insurance, and also to ensure sufficient quantities were available, a second order was placed on a well-known electrical firm. The results were most interesting. The ex-carpet factory in Scotland did what was asked without asking questions and produced a first-class product while the firm with the electric motor background tried to change everything to 'their way', resulting in a number of problems. This was borne out at the Director's meeting held on January 14th 1943, when it was reported that 'the production of electric airscrews had reached the stage where it was in excess of aircraft build, although motor delivery from two large electrical concerns was not coming up to expectation'.

Rotol and its subsidiaries must have become extremely proficient in the manufacture of electric motors, because two years later, in July 1944, at a meeting with the Rotol Directors, Coverley reported an enquiry for 250,000 motors for refrigerators. The first order would be for 50,000, at the rate of 1,000 per week – after the war was over. As events were to prove, Rotol did not pursue this enquiry, where the question of manufacturing them to a commercial cost would have presented considerable difficulties, as everything connected with aviation is usually expensive, because of the high manufacturing quality and inspection requirements essential to safety. In the years that followed, several non-aircraft projects were to founder on this vexed question of high costs.

The Rotol electric airscrew was only fitted to multi-engined aircraft, where its slightly slower rate of pitch changing was not considered detrimental to the aircraft's overall performance. With the passage of time, the advantages of the electric airscrew, particularly its ability to feather the blades, were overtaken by the continued advance of the hydraulic version and this meant its eventual disappearance towards the end of 1943.

Propellers of all types were under constant development at Rotol throughout the war, but production of existing types as replacements for squadron losses continued. This is presumably the reason for Rotol continuing to manufacture magnesium blades, in parallel with the wooden variety, until May 1942 when the Directors agreed to discontinue its use for the production of airscrew blades. Later in the same year, because of reported blade failures on the Halifax, steps were taken to replace existing magnesium blades with wooden ones.

Fedden, as Chairman of the Royal Aeronautical Society's Advisory Committee, had been requested by the MAP in 1942 to undertake an investigation into standardising the methods of operation for propellers for engines of around 2000–2500 hp and L.G. Fairhurst was co-opted on the committee. At this time LGF was Chairman of the SBAC Propeller Panel and therefore in a prime position to contribute to the discussions. Some years later, following discussions between Rotol and De Havilland, LGF once again found himself chairman of another SBAC committee looking into the question of propeller performance. The need for a common method of calculating the aerodynamic qualities of future propellers arose from the vast amount of data which was being supplied by the National Physical Laboratory from their wind tunnel research. After a lengthy period of discussion, sometimes involving rather contentious arguments between the firms taking part, a brochure was finally agreed. This proved to be a great success, with a world-wide circulation in several languages to various aircraft companies.

At the start of hostilities, the RAF fighters fitted with Merlin engines had fixed pitch two-blade airscrews but three-bladed constant-speed airscrews soon became standard for both fighters and bombers. As the war progressed, aero engine manufacturers steadily increased their horse power, and in order to cater for this the number of blades also increased, first to four, then five and finally six blades, this latter being a contra-rotating propeller.

The Spitfire Mk XIV, considered by many pilots to be the best of the breed, was powered by a Rolls-Royce Griffon engine, driving a five-bladed propeller. A 50 hour hangar type-tested propeller was completed at the end of 1942 and despatched to Supermarine for flight trials. The decision to change from four to five blades, specifically for the new Griffon engine, had been taken less than a year earlier by the Rotol Design Office, which is an indication of how quickly major changes could be implemented in times of need. The first production Spitfire Mk XIV flew in October 1943 and squadron deliveries began in January 1944, in plenty of time to help to counter the German V1 Flying Bomb attacks that started in the summer and for which this mark of Spitfire, with its greatly improved performance, was eminently suited.

Alongside the five-blader in the test programme, but a great deal more complicated, was the contra-rotating propeller. This had two banks of three blades rotating in opposing directions and had originally been intended to be fitted to the later marks of Spitfire. Because of its long gestation period it ended up on the navalised version of the Spitfire, the Seafire Mks 46 and 47, serving with the Fleet Air Arm, where the propeller's main characteristic, elimination of torque, was a blessing to many pilots, particularly when landing or taking off from the deck of an aircraft carrier in heaving seas. The contra-prop had more than its fair share of teething troubles and it was some while before production began, by which time the war was over. Just under 300 were produced for the Seafire installation and other contra-rotating propellers followed as the 1940's drew to a close, a handful for the Seafang – the naval version of the Spiteful – and a small order for an aircraft with twin Merlin engines, the Short Sturgeon. In the 1950's Rotol went on to develop further contra-props, details of which will be found in a later chapter.

In order to provide adequate cooling for the Hercules double-bank radial engines, particularly when the engine was being run-up on the ground prior to take-off or climbing under full power, Rotol developed the use of an engine cooling fan. This was a multi-bladed fan, positioned between the leading edge of the propeller cowl and the trailing edge of the spinner. The fan was driven at propeller speed, mounted on and driven by the spinner and was largely used in conjunction with the De Havilland Hydromatic Propeller, on hundreds of aircraft powered with the Bristol Hercules. Production of engine cooling fans began at Rotol's Scottish factory towards the end of 1942, when Blackwood Morton accepted a contract for 275 per week and another factory at Hunsworth in Yorkshire, originally taken over to make metal blades, undertook to produce a similar number. By July 1943, Ministry contracts for more than 23,000 fans had been received.

Because of the increased volume of work being undertaken at Rotol, a need to centralise the technical staff became apparent in 1942. The maintenance of scattered departments in line with the original dispersal policy, was no longer quite so important and in any case had given rise to a considerable amount of inconvenience. There was insufficient accommodation at Rotol for the move to take place and this led to the decision to build a new office block. The site chosen for the Technical Offices lay alongside the original offices, fronting the main road between Cheltenham and Gloucester.

The austere brick building, commonly known as 'the Tech. Block', was constructed during the latter months of 1942, at a total cost of £31,000, paid for by the Ministry of Aircraft Production, who also supplied the furniture. It was ready for occupation on January 17th 1943 and the Main Drawing Office and a few executive personnel were transferred from the 'White Block' (although at this time it was far from white, being covered with green and black camouflage paint) to the upper floor of the building and the Stress and Performance Offices moved to the ground floor. A number of employees from outlying offices in Cheltenham and Gloucester were also recalled.

In 1943, the set-up of the Technical Department was: Chief Engineer – L.G. Fairhurst, Chief Designer – E. Danvers (always known as 'Dan') and Chief Draughtsman – Frank Blair. The Main Drawing Office comprised a number of detailing sections, each of whom turned the original designs – or 'schemes' as they were known – produced by the Project Design Office, into individual working drawings which the machine shops could use. Section leaders, all with long experience in the drawing office, controlled each of these detailing sections as follows: Electrics – Ted Morris, Blade Design – Frank Bonsall, Hydraulics – John Boast.

The Stress Office came under the control of Ken Thatcher, with L.B. Jones as Stress Section Leader and D.V. Wordsworth as the Performance Section Leader. The Checking Section in the DO did not appear until a little later, when it was staffed by mainly senior men who were employed to (hopefully) pick up any errors which might have slipped through.

The Gearbox Section remained at Imperial House in Cheltenham until moving back to Main Works at the end of the war. The Project Design Section continued to occupy the topmost office over the Spinner Factory in Cheltenham but in time, it too joined the Main DO on the upper floor of the Technical Block, where it became truly 'the back room', because of its position at the very end of the offices.

John Boast, a slow, thoughtful individual, had taken over the Hydraulics Section in 1942 from W.A. (Wes) Tattersall and will be remembered by many of the older employees as someone who didn't make up his mind in a hurry, carefully considering all the options before coming to a decision. He had a shock of black hair and although John's hair is now snow white it is still as thick as ever. He started in the original Rotol drawing office in September 1938, working with

Jim Finch on the first hydraulic governor units. When I joined the Main DO at the end of 1952, Eric Stockdale put me into John's care and I will always be grateful for the time he took to give me such a good grounding in the intricacies of the Rotol system. My first taste of the DO was detailing some of the parts for the Victor Fuel Flow Proportioner but eventually I was to concentrate on the controller units which were part of the many varied Dart propellers.

The Production Section in the Main DO in 1943 was under Eric Stockdale, and as an illustration of the more rigid code that predominated in those years, Tony Smith – the son of Cliff Smith, one of the first employees – had just joined the section after finishing his apprenticeship and was sent home one day because he had come to the office without a tie! The wholly female Tracing Section was also entrusted to Eric's care, which because it was company policy to trace every production drawing onto linen, needed a large number of lady tracers. Original drawings were on paper which could be easily damaged if not carefully handled, and with the number of modifications that were sometimes necessary and the consequent extra printing that ensued, this could not be guaranteed. With the increasing use of plastic film ousting paper in the 1960's the need for tracers disappeared, to the regret of most draughtsmen. (A good many marriages were formed from the union of tracers and draughtsmen – the author and his wife being one example – and this meant a fairly high rate of turn-over in tracers.) J.H.M. Parker – always known as Mr Parker because he was a good deal older than everyone else – one of the original employees in 1937, was responsible for checking the tracings.

Before the war, a draughtsmen was required to design the job completely, including stressing, costing, weight estimating and planning (to ensure the job could be made on the company's machines). After the original design was approved, he then drew the individual manufacturing drawings from which the machine shops made the various parts. In short, he did everything that needed to be done from start to finish, except producing the component. In case there should be any suggestion of male chauvinism in this paragraph, it ought to be noted that on rare occasions draughtswomen were used in the drawing office. Usually they had been tracers with more than the normal aptitude for draughting, although it has to be said that draughtswomen who could do as good a job as a man have always been few and far between, with some notable exceptions, and the picture has not altered much in the last fifty years.

With the war, because of the chronic shortage of skilled men to fullfil the needs of the times, the process just described was changed. The Ministry of Labour introduced a training scheme for anyone who showed the least aptitude for wielding a pencil, in order to turn out much-needed draughtsmen, and the whole design process was broken down to enable individual skills to be used. This was the beginning of the present system whereby the design and manufacturing processes are farmed out to various sections and departments, each acting independently but in concert with one another. It has to be said that some of the original designers who can look back to those early days, do so with more than a touch of pride, as well they might.

Because of their specialised knowledge, members of the DO Electrical Section and the Development Department, were formed into the Signals Section of the

Home Guard. Their task was to provide telephone communication between the various Home Guard posts all round the factory and also across the airfield. Ken Thatcher was in charge of the Signals Section at the start but John Chilman took over later, with the rank of lieutenant. Years afterwards, John went to Dowty Fuel Systems to attend a meeting with the then Manager, N.H. Payne (not to be confused with N.S. Payne the Rotol Experimental Manager) and when he walked into the crowded room, Payne leapt to his feet, swung up a smart salute and shouted, 'Good morning, Sir'. It may have been instinctive, as he had been Chilman's sergeant in 1941.

The Signals Section, according to John Chilman, were notorious for always managing to avoid the normal drill parades, marching, and general 'spit and polish' considered to be so essential for military life. One day, the powers-that-be, in an endeavour to get some discipline into this section, ordered them to the nearby rifle range, on the far side of Staverton airfield, to take part in target practice. John Chilman had represented Essex in this sport before the war, a fact he did not reveal, but he did ask if he might use his own .22 rifle. This was agreed, as the Home Guard used specially adapted rifles firing .22 ammunition for target practice. When he tells this story, John's sense of modesty prevents him from dwelling at any length on the subsequent result of this event, but one can well imagine that someone who shot for his county must have shown considerable improvement on the normal Home Guard member!

A specially built concrete hut was used as the headquarters of the Signals Section, located in the area now used as the small Executive Car Park. According to John Chilman, the telephone switchboard in this hut was occasionally manned (if that is the correct word) by the female operators from the Rotol switchboard. This is the first time I have heard of any women serving with the Home Guard and it would certainly not have been countenanced by Captain Mainwaring of 'Dad's Army' fame!

Terry Elmes was among the Rotol men from the Technical Department who joined the LDV at the outbreak of war, and boasted that he was the only man in the Home Guard who was excused parades, because no steel helmet could be found that was large enough for his $7\frac{3}{4}$ size head! We will meet Terry when we consider the part played by Rotol in the marine business, in the meantime he provided the following story, with some additions from John Chilman:

> A large scale Home Guard exercises involved a mock attack on Cheltenham by a company of the Regular Army, with Cheltenham Home Guard acting the role of the defenders. The Signals Section were to take the part of Fifth Columnists, and Reg Streete and John Chilman made up some portable 'walkie-talkie' sets for their fellow conspirators. On the day in question, the tanks of the attacking force were approaching Cheltenham from the direction of Dowdeswell, on the main A40 road, when they were met by a member of the 'Fifth Column' who guided them down country lanes to avoid the road blocks which had been set up. In less than half an hour, others of the 'enemy forces', led by Derek Warburton-Brown of the Signals Section, and wearing a spurious version of the Home Guard uniform, bluffed their way into the defender's HQ (the former Shirer and Lance building in the centre of Cheltenham) and captured the whole staff. The Commanding Officer protested in somewhat forceful terms at the underhand manner by which his forces had been

defeated and demanded that an inquiry be held. At the subsequent hearing, the action taken by the 'Fifth Columnists' was not only upheld, but they were also praised for their ingenuity. There was really no excuse for the failure of the Cheltenham Home Guard as the Signals Section had discussed their plans on the previous evening, using the walkie-talkie sets, without appreciating that everything they said was being transmitted on the BBC radio frequencies, for all to hear! Shortly thereafter, the use of these 'home-made' sets was forbidden by the authorities.

The Main Drawing Office were concerned with the component drawings, but each of these had to be converted into a finished part as indicated in the earlier paragraphs, which was the responsibility of the Rotol Production Department. This encompassed a vast empire throughout the works, ranging from the actual machine and assembly shops through planning, progress, stores, inspection, heat treatment, painting and plating – the list is almost endless.

Morris James came to Rotol from his home in South Wales for an interview with Cyril Taylor (Personnel Manager) and Bill Nottingham (Chief Tool Designer) in March 1943. He joined the Jig and Tool DO under Cliff Faulkner, where his Section Leader was Jim Disney, looking after gauges and fixtures and was considered one of the fastest draughtsmen in the department, by his contemporaries. Morris had previously been in a small engineering company, with only four draughtsmen, and was almost overwhelmed by the size of the Rotol office, which included a large Planning Section. As the years went by he he slowly climbed the promotion ladder, taking over the section from Jim Disney and eventually reached the position of Chief Tool Designer, when Cliff Faulkner went to the Training Department and Bill Nottingham was made Chief Planning Engineer in control of the Production Planning Department.

Four years after the end of the war, the Government ordered two prototypes of the Vickers Valiant, the first of the RAF 'V' bombers, and Morris James was given charge of a special project, as part of Rotol's contribution to the 'Super-Priority Aircraft Plan', which had been instituted. Nothing was to be allowed to hold up the production of parts required for this aircraft and as Rotol and British Messier were making the undercarriage and flap gearboxes, as well as part of the hydraulic system, Morris had to ensure that all the special tools and gauges were available. If Rotol did not possess the required special tooling, he had the authority to make a direct approach to other engineering companies where the tools were known to exist, which might be anywhere in the British Isles, in order to use their facilities. The Valiant first flew in May 1951 and in October 1956 was the first British aircraft to drop an atomic bomb, on a test range in Australia.

In later years, the changing character of new products that flowed through the Production Planning and Machine Shops, coupled with the increased activity of British Messier, brought different tooling concepts and Morris was asked to run a Special Products Section in the Jig and Tool Department with Ted Hughes as Section Leader. Over the subsequent months they looked into deep-hole boring, pressure welding and many other new methods, including tungsten carbide specialised tooling for tube boring, using knowledge acquired from German industry at the end of the war. After being with Rotol for 16 years, Morris left for other fields, but returned on a weekly basis for many years as Technical Sales Consultant, covering tungsten carbide and diamond products.

In the early part of 1943, an interesting but somewhat hazy event occurred and apart from the account in D.M. Hill's book *Ladies may now leave their machines*, not much is known. She tells of a 'revolt' by the workers, culminating with a mass meeting in the canteen, which ended in a call for action by higher authority. The main grievance was not about pay or conditions but of the concern being felt for the supposed inadequate use of the machine shops, and it seems to have centred around some machines being idle at a time when the Government were urging for a greater effort from the workers. The obvious feelings of disquiet evidenced by the workforce was sufficiently alarming for the management to ask Sir Stafford Cripps, the Minister of Aircraft Production, to come down and address them in order to allay their fears. He arrived on March 6th 1943 accompanied by Lady Cripps and several officials from the MAP including the Director of Public Relations and also the Regional Personnel Officer. In addition there were a great number of Rotol management, making an impressive total of around eighteen in the party.

Although no reason for the visit is given, an account in *The Airscrew* gives some idea of what was said by the Minister, when he addressed a vast crowd of workers in the canteen. He dealt with 'the difficulties of maintaining a balance between an even flow of production and at the same time continually modify that which you are producing,' and went on to point out that 'for security reasons one cannot always explain the next development in the air battle, and that which appears wasteful of production is a responsibility which rested on the Minister's shoulders.' Sir Stafford must have spoken at some length on this and other related topics and ended with offering thanks from the Government for the prolonged efforts of the Rotol workers. The explanations were presumably accepted, indeed there was little alternative, and after the party had completed a tour of the works and departed, life went on as before and 'nothing further was heard of the Rotol revolution', according to Miss Hill.

One story concerning Sir Stafford Cripps happened when he visited the Flight Test Department, where he encountered one of the riggers doping a patch onto an Airspeed Oxford and singing at the top of his voice. Always keen to show the workers that he was interested in their activities, Sir Stafford asked him what he was doing. 'Buggered if I know, mate' came the casual reply as the fitter returned to his dope and the rest of the song. Sir Stafford shook his head and continued on his way, perhaps reflecting on the instructions of the Official Secrets Act which forebade any employee passing on details of his work to an outsider.

Throughout the war, it was quite common for the Flight Test Department to have more than a dozen aircraft on test at one time, the busiest period being 1945 when 32 aircraft were on charge over the year. By the end of the war, the Flight Test Department, by virtue of the large number of aircraft it was testing, became known as 'Greensted's Air Force' by those in the local area.

The work done by the department was of a varied nature. Tests involved the installation of the propellers or other equipment, the test equipment itself and any hydraulic piping or electrical wiring needed for its operation, and connections to on-board power supply. The installation of strain gauging, automatic observers and recorders with their instrumentation, often meant removal of radios or operational equipment. When the flying programme was completed, test equipment had to be taken out and the aircraft restored to its original condition for return to the manufacturer or the appropriate RAF unit. All this meant a heavy work load and it was decided to introduce female labour into the Department. Suitable candidates were chosen from the applicants and Ted Edwards recalled that their appearance on the scene caused quite a stir among his colleagues. The Flight Test Department was not alone in feeling the impact of the sudden arrival of several young ladies into what had been a completely male-orientated world. Many times, during the research for this book, the author was told of the friendships, liaisons and romances which developed at Rotol during the wartime years, when a large part of the work force comprised young and unattached women. In the interest of truth it has to be recorded that sometimes the romances involved married women as well. However, interesting though this may be, let us return to the flight testing which continued despite the distractions.

Other units based at Staverton during the war were the Flight Test Department of the Gloster Aircraft Company and training units of the Royal Air Force. The Rotol test pilots were sometime called upon to fly other aircraft, in addition to those operated by Rotol, including those from Flight Refuelling and also Folland Aircraft. This latter company produced a flying test bed to be powered by the Bristol Centaurus, which was virtually a 'double-Hercules' giving over 3000 hp and Michael Daunt, a Gloster Aircraft test pilot, began the test flight with a Hercules engine installed. The wooden-structured aircraft broke up during diving trials and Daunt took to his parachute and landed safely near the Officers Mess. (This building is still in use by the firm of Wynnstruments at the time of writing.) He went into the bar, looking somewhat the worse for wear, demanded a large Scotch and declared that he was not having anything to do with the further trials. Squadron Leader Dickson was elected to continue with testing the second prototype and in due course he gingerly put the aircraft through its initial test programme. On June 20th 1943 the huge Centaurus was installed but on his second flight the constant-speed unit malfunctioned and the propeller blades moved into the feathered position. With a 'dead' engine, Dickson skillfully managed to put the Folland down on the grass alongside the runway at Staverton.

Testing was also carried out at Staverton on the Dowty undercarriage fitted to the Halifax bomber. Dickson recalled one exercise that required 100 landings and he carried out 15 to 20 in quick succession, when the sight of this large aircraft performing rapid circuits at low-altitude caused a great deal of local interest. Dowty's representative, Fountain-Barber tried to persuade them to allow him to be strapped to the undercarriage in order to observe the impact on landing, but was fortunately dissuaded!

During this period, Dowty's were experimenting with hydraulic suspension for a motor car and Fountain-Barber's was so equipped. One evening he insisted that Dickson and some of the others made a circuit of the perimeter track to demonstrate its handling. Let Dickson tell the story:

> We each took a turn, driving leisurely around the aerodrome, but he complained we were too slow. Against my better judgement he persuaded me to take a quick circuit, but taking the final bend much too fast, the car turned over several times and

he was thrown out with severe head injuries. Luckily I remained inside. During the afternoon he had collected several punnets of raspberries from Evesham and when my wife hurried over from the Officer's Mess she saw me covered in this fruit, which she at first mistook for blood. No more car testing occurred.

Readers may be interested to learn that Rotol played a part in the manufacture of the weapon invented by Barnes Wallis for the destruction of the German dams. Although I can find no documentary evidence, I have it on good authority that Rotol made the casings for the bomb, designed to skip along the surface of the reservoir and crawl down the face of the dam wall, before exploding. As Wallis did not get the go-ahead for the weapon until February 1943 and the attack by a special squadron of Lancaster bombers took place in May, three months later, we can be fairly certain the casings came through the factory around that time. An example of one of the unused bombs may be seen at the Imperial War Museum airfield at Duxford, near Cambridge.

With the exception perhaps of the Research Department, which will be covered in a later chapter, no aspect of the company's activities claimed so little attention as that of KBO and its allied sections. Keynsham Bank, a large house on the London Road leading out of Cheltenham, was taken over by the Rotol Accounts and Cost Departments who were moved there at the beginning of the war. M.H. Stephenson, one of the very first employees in 1937, had charge of the Cost Department and the well-liked and highly respected Chief Accountant was W.A. (Bill) Barr. The house itself was large and rambling, occupying several floors, with all the rooms turned into offices. One unusual feature in the garden was a magnificent mulberry tree.

Keynsham Bank Office, or KBO as it was known in official parlance, was described in *The Airscrew* as being shrouded in mystery, and so it probably was to those who did not work there. Financial matters tend to have this effect on some people, if the author is any judge. It was certainly a pleasant place to work and had the advantage of being close to the shopping centre in Cheltenham. Mrs Sylvia Weller (née Beasley), was transferred with the rest of her colleagues from Main Works and remembered the years spent there with great affection. There was a high degree of mechanised accountancy (Sylvia herself operated a comptometer) and although the staff was small, a high proportion were married women and part time workers.

In common with all the dispersed departments, Keynsham Bank had its own canteen, run by a lady of European extraction named Mrs Muller and one of her specialities was a kind of goulash stew. Sylvia had good cause to remember a particular lunch-time at Keynsham Bank. She had been down to the kitchen area to collect a bowl of stew and was taking it back to her office on the top floor when she tripped over one of the steps and shot the whole lot across the landing leading to Mr Barr's office. He put his head round the door, saw the mess and enquired if she had hurt herself. The only thing that was hurt was Sylvia's dignity but a notice soon went around Keynsham Bank, forbidding anyone to eat meals anywhere except in the room set aside as the canteen.

'Fire-watching' was not, as the name implies, a matter of simply standing around as an interested spectator, while a conflagration took place. As part of each able-bodied person's ARP duties, two people had to spend the whole night in their offices, thus ensuring that if the building was struck by bombing and set on fire, somebody would be on hand to put out the flames before they could take hold. The Government had put this into effect following the disastrous bombing raids by the German Air Force soon after Christmas 1940, when a large part of London had been set ablaze by incendiary bombs, many of them started in locked offices. To learn how to cope with the effects of fires and of incendiary bombs in particular, which could be rendered harmless if they were caught early enough, involved a certain amount of training. Sylvia, although only a young teenager at the time, had to take her share of fire-watching, in company with another female, although thankfully, she never had to put her training into practice. Another Rotol employee who spent many leisure hours at Main Works in fire-watching was L.G. Fairhurst, during which time he produced a large number of scale model aeroplanes from scratch. This was in the days before plastic kits were heard of and one started with a plan, a knife and a piece of wood, and the centre pages of a copy of *The Airscrew* is devoted to a selection of models made by LGF.

Harry Wood and Basil Herbert both worked in the Cost Department at Keynsham Bank, and were scheduled to be the official fire-watchers on one occasion. Before beginning their duty, they decided to have a drink in the local hostelry and on returning to the now deserted offices, felt rather peckish. Scouting round the canteen they came upon a jar in which were some scraps of meat and they proceeded to have a 'fry-up' with some other bits and pieces. The next day they had to face the wrath of the Canteen Manageress, Mrs Muller, who accused them of stealing the scraps she had been saving for her cat, rescued from the food left on the plates by the diners.

The Rotol guest house has already figured in this narrative, but the next tale, involving a foreign visitor, was recounted by LGF and may not be as well known as the classic story of the marine shaft, which is told in a later chapter. He received a phone call from the Chinese Embassy in London, asking if the company would allow their Air Attaché to pay a visit on February 20th 1943, to discuss various projects and the possibility of licence-building the Rotol airscrews. He was invited to stay overnight at Battledown Manor, the Rotol guest house. RHC suggested that in order to make their foreign visitor feel at home, the lounge should be decorated in Chinese fashion and one of his staff was sent to Sothebys to purchase something suitable. A porcelain figure of a prancing horse was chosen and set up in a prominent position. On the evening in question, Major Huan Pen Young was introduced to the few guests and the evening began. After some time RHC drew LGF to one side and asked if any mention had been made of the figure. On receiving a negative reply he urged him to casually bring up the matter, in order to get something from the investment. Being rather lost in matters concerning the world of art, LGF cast around for some inspiration and eventually came up with a suitable question. 'Major, we've had this porcelain figure for some time and wondered if you could give us some idea as to what dynasty it belonged.' The reply was a blow to his listeners. 'I'm afraid I can't help you, all my life has been spent in the United States – I've never been to China.' The porcelain figure was returned to the sales rooms the next day and nothing further was heard of it. Perhaps it has already had a second moment of glory on the popular television programme, 'The Antiques Road Show'.

As a postscript to this tale, it is interesting to find that as a result of these discussions, Chinese officials were negotiating a contract to make Rotol propellers under licence when the war ended, but because of the Communist take-over of power the plan never got any further.

HM Queen Mary paid her second visit to Rotol on the afternoon of May 3rd 1943, accompanied by HRH The Princess Royal and her husband, the Hon Gerald Lascelles, Lady Constance Gaskell-Milne and Major Wickham. They were met in the Reception Hall at Main Works by Mr George Spencer, Regional Controller for MAP and introduced to the various members of the Rotol management. Following the signing of the visitors book, the Royal party made a tour of the works. Tea was served in the Staff Canteen and Miss Kathleen Berry, a charge hand Inspector, presented a bouquet of flowers to Her Majesty. The visit concluded with a look around the Flight Shed, where Bryan Greensted showed the party all the aircraft which were in the hangar. This part of the visit received the greatest attention by the Hon Gerald Lascelles, who was extremely interested in aircraft.

At one point the Queen was being shown the Halifax which obviously impressed her because of its size, when she turned to Squadron Leader Dickson, one of the Rotol test pilots, and pointing with her parasol enquired the weight of the wheel. Dickson said that he had no idea, a frank reply which possibly amused her. In his words, 'It was possible to notice a slight twitch to the mouth which on a lady of more humble origin might be construed as a grin'.

One story has come down from the visit, which is probably apocryphal but could so easily be true. It was customary for the Queen to have a few words with several people on her tour, and stopping alongside one of the women operators she asked, 'And what are you making, my dear?' 'About time and a half, your majesty', the girl replied.

In the Experimental Shop, the Queen was making her way down the line of machines when to everyones surprise, out stepped Eric Saxton, looking very smart in his Home Guard Uniform, crashed his boots together and whipped up a salute which would have done credit to the Brigade of Guards. 'May I have your autograph please, your Majesty', said Eric, offering her his book. A moment of embarrassed silence followed and some confusion while a pen was found and the Queen graciously signed her name. Don Williams took a number of photographs showing the Queen flanked by ladies from the various sections of the factory and offices and they were subsequently printed in the next edition of *The Airscrew*. The Royal visit seems to have been a bright spot in a long dreary war, and RHC even used a photograph of himself with the Queen, as his personal Christmas card. As Eileen Hunt, who remembered the visit clearly, remarked, 'Queen Mary showed us how elegant we all might be, given the clothing coupons and the make-up . . .'

VIP's were not confined to merely visiting the Main Works at Rotol, as the dispersal factories also played host to some illustrious personages. The Princess Royal, once more accompanied by her husband, the Hon Gerald Lascelles, was shown round the factory at Ossett on July 16th 1943. Such an important occasion needed the presence of a strong contingent from Main Works, including R.H. Coverley, R. Stammers and Miss D. Ashworth, the Women's Welfare Officer.

One of the decisions taken by the company in 1943, was to give rise to some heated discussion over forty years later, when my research was in its early stages. This came about when it was proposed to drop the word 'Airscrews' from the name of the company. There is a folk legend in the company, that the change came about because of the confusion between 'aircrew' and 'airscrews', resulting on at least one occasion in the sending of several aircrew to an RAF station, in place of a set of airscrews. Although this is a fascinating yarn, I was frankly sceptical and veered towards the opinion that the real reason lay in the fact that Rotol had moved from being solely a propeller-making company by this time. It made sense to drop the word 'Airscrews' both on the grounds of sound business judgement and possibly because the word itself had become somewhat out of date. (There was some mild controversy in the aviation press of the period, as to whether an airscrew was a propeller.) When I made a search through periodicals around the end of 1943, I soon found my doubts were justified.

The change of name was announced by Coverley at an official luncheon in London at the end of October 1943 (further details of which will be found in the following chapter), when he pointed out that the company had considerably expanded its activities in the seven years which had passed since its inception. An announcement must have been made to the press about this time, because in the copy of *The Aeroplane* of November 12th 1943, hidden away among its back pages, was a five line item which reads: 'Rotol Airscrews Ltd have changed the name of the company to Rotol Ltd. The company formed jointly by Rolls-Royce Ltd. and the Bristol Aeroplane Co Ltd for the purpose of airscrew development, has widened the scope of its operations.' It had taken some time for this to come to fruition, as the decision had been made by the Board of Directors, as far back as their meeting in July of that year, when the secretary was instructed to proceed with the necessary formalities. Although the minutes of the meeting do not give any reason for the suggested change, it seems logical to assume that it was because of that put forward in the press. No written evidence to support the supposed mix-up between 'aircrew' and 'airscrew' has come to light but even at this late date someone may have knowledge of such a directive in the old archives. In the absence of further evidence, I am convinced that the explanation put forward is as close to the truth as we may now get, although it is certain to be rejected by many of the old hands.

The question of providing living accomodation for the many workers transferred to Rotol by the Minister of Labour from all over the country, was an important topic and in order to ensure this was properly handled, the Welfare Society were involved. Some workers were billeted with local families, at least for a time, but this arrangement could not be expected to cater for large numbers. The solution, both here and in many other parts of the country, was the setting up of a special hostel. The company had always rented out a few houses in the district to selected employees and in charge of the billeting arrangements was F.C. Bunn. He was also the Welfare Officer, and acted as secretary on both the Entertainments and Bus Committees. When the Rotol hostel was built, comprising a number of huts on some waste ground near The Plough on the site now occupied by the Staverton Technology Park, this was added to his other duties. It was completed sometime in 1943 and immediately occupied by around 300 Rotol and 200 Dowty Equipment workers. A number of employees who were interviewed for this book had their first taste of Gloucestershire at the Rotol hostel. Some years later, the author spent the month of January 1953 there, and can vouch for its being fairly spartan but – for anyone used to service billets – it was reasonably comfortable, and the rabbit hot-pot was particularly delicious!

At a meeting of the Board of Directors in November 1943, a rather unusual incident concerning Pobjoy and Coverley occurred. Sir Stanley White casually enquired why Rotol had issued a brochure indicating that they were considering entering the field of aircraft manufacture, with either helicopters or autogiros. Coverley was somewhat taken aback, being unaware of the brochure, but promised to look into the matter. He returned from his enquiries, with a denial from anyone having knowledge of the matter, but White produced a copy of the brochure from his pocket, with Pobjoy's name on it. Coverley then retired once again to confront Pobjoy who then confessed to the whole business. It transpired that Mr H. Bolas had been appointed to the post of Auxiliaries Project Engineer, to look into the question of Rotol products being used in helicopters and had presumably put out the article in question, but why Pobjoy did not elect to obtain Coverley's approval before issuing it to the Directors is not known. Pobjoy's actions would certainly not have found favour with either the Controller or the Directors and although the outcome may have been nothing more than 'a storm in a teacup', ultimately it must have affected Pobjoy's career at Rotol.

As already recorded, the discussions between the management at Hordern-Richmond and Rotol concerning a proposed take-over in 1941 had proved abortive, for a variety of reasons. The Directors had always been concious that although Rotol was in the forefront of wood blade airscrews, they did not control the basic product from which the blades were made but now they believed the time was right for remedying this state of affairs. Talks were resumed in 1943 and a satisfactory agreement was finally reached in August, when Rotol took a controlling interest in Hordern-Richmond, with the purchase of 134,200 of the 5/- Ordinary Shares. As Hydulignum blades could be made with thinner cross-sections, giving a worthwhile increase in performance, they were preferred to either the Jablo or Weybridge types as it enabled them to be fitted to the new generation of fighters where there was a demand for higher speeds.

No matter which type of wood blade was used, they all required to be covered with a protective film, in order to withstand the severe conditions under which they had to operate. These coverings came with a bewildering range of names; 'Rotoloid' was the trade name for a cellulose-nitrate sheet, developed by Rotol, while the Airscrew Company used 'Rayoid', a cellulose-acetate sheet. Both types were .04″ thick and could be applied to either Jablo or Hydulignum blades. 'Cristofin' was a Hordern-Richmond process, used on Hydulignum blades only. Jablo covering consisted of a bronze gauze wrapped around the blade and sprayed with several coats of synthetic resin enamel and was used on Jablo blades only. Finally, Schwarz covering was a cellulose acetate sheet, reinforced with fabric, used on the Weybridge blades. This brief summary has inevitably simplified the processes concerning protective blade covering, which were far more complex in reality. One further addition to the wood blades remained. A special form of armoured steel leading-edge sheath was successfully developed and patented by Rotol in the early part of the war. Its purpose was to prevent the wood blades from shattering if they flew into balloon-barrage cables – not always those belonging to the enemy – and was fitted to multi-engined aircraft.

One important point has not been aired when considering the vast number of Rotol airscrews being produced, that of cost. It would not be feasible to go very deeply into the question (although a few prices are quoted in the appendix section) but of course the subject came up for discussion at many of the Director's Meetings. In May 1942, Sir Roy Fedden tabled a report of a meeting he had attended, concerning standardisation of propeller types, at which it was apparent that the Rotol propeller was considerably higher priced than the De Havilland product. The difference was almost entirely accounted for by the higher cost of wood blades as compared with Duralumin, and subsequently steps were taken in an attempt to improve the situation. Sidgreaves made a fair statement when he pointed out that to make a true comparison of costs, due regard should be taken of technical, weight and repair advantages of wood versus metal.

The wood blades that Rotol had pioneered with such success were always high on the requirement list for new engines and aircraft – or at least on the improved versions which regularly appeared as the war went on – but the question of metal blades for the future had never been far away. In April 1943, the MAP sanctioned a scheme put forward by Rotol, involving the re-organisation of No 1 Machine Shop, and the removal of the Metal Blade Shop to Ossett in order to cater for the anticipated production and assembly of five-bladed and contra-rotating propellers. There were several reasons why this decision was made: firstly, more space was needed at Main Works – as it always was – and secondly, the company had received an order for propellers fitted with Duralumin blades for the Centaurus-engined Hawker Tempest Mk II where the extra weight provided by metal blades was needed to maintain the aircraft's centre of gravity in the required position.

The Metal Blade Shop at Main Works was transferred to Ossett under the watchful eyes of the Works Engineers, at a cost of £8,600, paid for by the Ministry and the whole process was accomplished in six weeks. For the first time for many years, Rotol began to receive substantial allocations of Duralumin forgings from High Duty Alloys of Redditch and although it was only a small part of the overall production, this appears to have been the beginning of the end as regards wood blades. Ossett commenced metal blade production in August 1943 in a new shop known to the rest of the factory as the 'Iron Blade Shop', in order to distinguish it from the 'Wood Blade Shop'. Eventually, over 1,250 propellers for the Tempest Mk II, Mk V and Mk VI were produced.

The Supervisor in the metal blade shop at Ossett was E.P. (Ted) Marshall, who will be remembered by many people in Churchdown, where he served over thirty years on the Parish Council. Ted was apprenticed with Fielding and Platt in the late 1920's, then after a spell with R.A. Lister at Dursley, he joined Rotol in June 1937. He started in High Orchard Street as a metal blade machinist under Reg Barrell, later transferring to the Experimental Shop as Night Shift Foreman. When the war started, Ted's experience was needed in the new factories in the North of England, but he also spent some time with the bomber squadrons of the Royal Canadian Air Force in Yorkshire as a technical advisor and trouble-shooter. Every third week he was allowed weekend leave to travel home to the house in Parton Road, which he still occupies at the time of writing.

Rotol Rugby Football Team 1940/41. *Left to right*, back row: L.C. Rowles, E. Martin, J. Strain, B. Hardiker, B. Teague, T. Price, W. Perry, W. Jones, T. Walker, J. Smith, S. Bartlett, G. Richardson. Middle row: T. Whitney, C. Bloxsome, G. Oakley, W. Lewis, W. Warner (Captain), M. Kleinman, J. Tandy, J. Pickford, E. Jones. Front row: S. Walden, J. Griffiths.

Ladies hockey team in 1943. *Left to right, back row:* J. Gapp, J. Harris, M. Nelson, M. Webb, M. Fry, Mrs. Orritt. *Front row:* P. Clutterbuck, G. Kilpin, L. Long, J. Quilter, I. Belcher

Bryan Greensted and Jack Hall in front of a Whitley bomber, with the airscrew feathered. Much of the Rotol research work was done on this aircraft.

A Spitfire Mk IX fitted with an experimental Rotol contra-propeller. Jack Hall was killed when flying a similar aircraft in July 1944.

The first production Spitfire I fitted with an early 35° airscrew in 1939. *Via Jack Meaden*

The Wellington VI was a high altitude version with a pressurised cabin in the forward fuselage. DR475 was used by Rotol FTD from June 1942 to July 1944. *Via Jack Meaden*

Westland Wyvern S4 strike fighter shows off its Rotol eight-blade contra-prop in February 1953. It became the last Rotol FTD aircraft to leave Moreton Valence in June 1954. *Via Jack Meaden*

A Royal visit – but work on a Fairey Barracuda Mk I goes on. Queen Mary at the Flight Test Department hangar on 3rd May 1943. *Left to right*: Bleaney, Stammers, Greensted, HRH, Coverley, the Princess Royal, Law, Cameron, Miss Ashworth, Miss Jackson, Dickson.

HRH and Olive Dyer (née Ledbury) in the Rotol Training Department. Noticing that Olive had shyly turned away the Queen commanded 'Look up missie' and Olive dutifully obeyed.

Rotol Home Guard, Bomb Disposal Squad, in 1942. Ben Harris (seated centre front) was in charge. Standing on the left is Jim Brindle with Stan Booth beside him.

Servicemen being shown the workings of a Rotol propeller in the Instruction School, 1943.

A Rotol romance. Denis Savory and Jean Hockley in 1946, some two years before their marriage. Jean worked in the Print Room under Harold Williams at the time of this photograph.

Two very competent ballroom dancers, Eileen Walker (Pobjoy's secretary) and Ivor Brooking (Project Designer), were persuaded to act as judges at a Rotol dance competition in 1942.

Forty four years separate these two photographs. John Hunt, an apprentice in 1944 (left) and just prior to his retirement from the Design Office in 1988 (right).

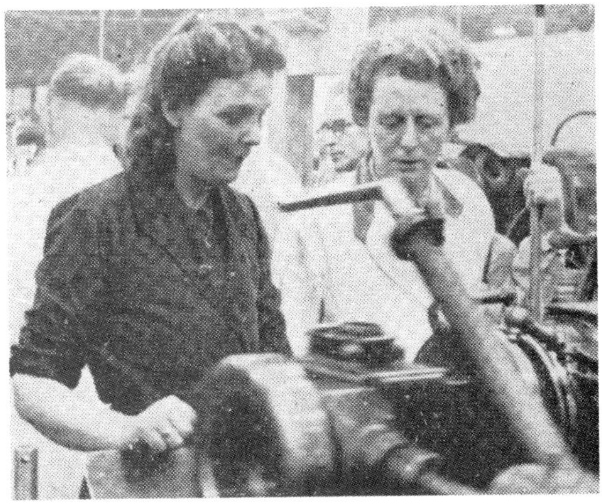

Mrs. Mary Moss, a Shop Stewardess (left) with Miss 'Mac' McGonigal who was in charge of all the ladies in both No 1 Machine Shop and the Assembly Shop in 1943.

Group Captain 'Sailor' Malan on a wartime visit to Rotol.

Roger Armstrong, Bronze Medal winner in 1943.

Sister Rees in charge of the Rotol Medical Department (taken in 1943).

A visit from Sir Stafford Cripps on March 6th 1943. The Rotol Controller, R.H. Coverley, is on the left.

Pobjoy and Coverley survey the Rotol AGP6 auxiliary generator which was fitted to the Short Shetland flying boat.

A general view of the Production Offices with the Jig and Tool Drawing Office on the right.

Awaiting the daily rush in the Rotol works canteen. The stage can just be made out in the far left corner.

Rotol Apprentices Guild AFC, League winners of Cheltenham II Division in 1946. Eddie Cowley is seated left. Reg Alexander is standing second from right.

The camouflaged 'White Block' is lit up on VE Day – May 8th 1945.

The wartime Experimental Shop before alteration to the New Tech. Block

Happy smiles from the Experimental Shop workers in 1948. Seated right is Charlie Beasant, Shop Foreman, next to Charlie Prior his deputy. Arthur Gunn (in white overalls) stands behind.

A recent photograph of a preserved Spitfire Mk II P7350 of the Battle of Britain Memorial Flight, with Rotol propeller. *Robert Rudhall*

Rotol propeller on the Trent Meteor. On test the blades were subsequently cropped to just under five feet diameter.

The Closing Months of the War

Towards the latter months of 1943, the vast production might of the United States began to be felt as great quantities of men and machines started to crowd into the British Isles, in preparation for the forthcoming invasion of Europe. The American propeller industry had gone to enormous lengths to increase its production capacity, converting factories from consumer products to propellers for its bomber and fighter aircraft. *Thrusting Forward*, the history of the Hamilton Standard Company, states that 75% of the propellers used by the Allies during the war were supplied by the United States, which increased its production from 10,000 per year in 1939 to 100,000 in 1943. One of the consequences of this immense strengthening by the Americans of the Allied air effort was a report from the Ministry of Aircraft Production, which gave the Rotol Directors a severe shock when they learned that the anticipated British propeller programme for 1945 would be halved.

All the major aviation companies in the country were affected in some way or other, but Rotol had cause to be particularly worried as due to the efforts of workers at dispersal factories and Main Works, their existing propeller programme was already ahead of schedule. Ways and means of countering this forthcoming reduction were considered, starting with a gradual transfer of work from the sub-contractors and a possible extension of the auxiliaries programme. The post-war situation was also beginning to feature in some of the management's discussions, and Rotol began to look for other products – not necessarily connected with aviation – which could keep the works and offices busy during the weeks and months that lay ahead.

The unease for Rotol's future was well placed. At the meeting in March 1944, the Directors were informed that turnover had dropped from £2,610,116 in 1943, to £1,568,951 in the same period for 1944. It was suggested that the Government should be made aware of the concern felt by aircraft manufacturers at the reductions of output in British factories.

Distinguished visitors continued to turn up at the factory throughout the later war years. Mr Lennox Boyd, the Parliamentary Secretary to the Minister of Aircraft Production, was shown round on February 1st 1944 and in April two Air Marshalls made an appearance, Sir Arthur Barratt, the C in C Technical Training Command on the 18th and Sir Ralph Sorley on the 20th. Sir Ralph had been a great believer in the Hurricane fighter in the bleak years before the war, and was eventually to become Managing Director of De Havilland Propellers in the 1950's.

Strange as it may seem, among all the developments being made with regard to multi-bladed propellers, each one an advance on the previous model, Rotol had one of its biggest successes on a far less ambitious product. Towards the end of

1943 and for some years afterwards, a two-bladed feathering propeller was produced for twin-engined training aircraft, replacing the original two-bladed fixed pitch version. The propeller was first tested on an Airspeed Oxford and although none were ordered for Oxford production, more than 3,250 were manufactured for the Avro Anson and its civil counterpart, the Avro XIX.

According to L.G. Fairhurst, there was a personal reason why Rotol wanted to fit feathering propellers to these aircraft. Stammers and LGF had taken off from Croydon in a twin-engined Oxford following a meeting in London, when one of the engines failed. He described the terrifying experience of watching the propeller, windmilling and perhaps churning up broken pieces in the engine, as the aircraft steadily lost height. Fortunately, the pilot made a successful landing in a field and LGF believes that may have been the moment when he decided that the Design Office should start work on a two-blade, constant-speed feathering propeller.

June 6th 1944 saw the beginning of the end of the Second World War, when the largest amphibious operation ever undertaken set sail from the British Isles, headed for the Normandy shore. Except for the people who watched the increased aerial activity as bombers, fighters and transport aircraft flew overhead, or had loved ones in the camps who were awaiting embarkation, this momentous occasion went largely unremarked at Rotol. All over the country there was probably a sense of relief, that after so many years of war, the waiting period was now over. Although success was by no means a foregone conclusion, there was light at the end of the tunnel.

Rotol experienced its share of fatalities throughout the war years, for one reason or another. The names of employees who had joined the armed forces were occasionally to be found in the pages of *The Airscrew* as having died while on active service and others were killed in accidents. Test flying ranks among one of the most dangerous occupations and it was on Saturday, July 29th 1944, that Ted Edwards was driving his car from the Main Works when he heard a loud cracking noise over his head and looked up to see the Rotol Spitfire Mk IX fitted with a contra-rotating propeller, go into a steep dive from a low altitude of perhaps 600 feet. The aircraft hit the ground in the direction of nearby Boddington and Ted made off as quickly as possible for the crash site. An army signals unit was stationed at the little hamlet of Barrow and on enquiring at the guard room he was told that the crash had occurred in the compound. Ted was unaware which of the three pilots had been flying that morning and showed his credentials to gain access to the scene of the crash. One look in the smashed cockpit showed him that Jack Hall had made his last test flight. The report submitted to the Ministry put the cause for the accident on the leading edge protection on one of the blades becoming loose, causing severe vibration leading to the break-up of the six-bladed contra-prop. The broken pieces struck the tail unit which then broke up in its turn. Jack Hall would have stood no chance at the low altitude at which he was flying. A modification to remove all metal leading edges from contra-props was put in hand at once. Squadron Leader Clive Kirkland, a former Halton apprentice and a contemporary of Bryan Greensted as a flying instructor at Air Service Training, replaced Jack Hall. A second fatality at the Flight Test Department occurred when Flt Lt Philmore, a serving RAF officer, crashed on February 18th 1946 when flying a Spitfire Mk 21 in bad weather conditions near Barnstaple in North Devon. He had been seconded to Rotol to test single-engined aircraft.

It was during the latter years of the war that the Auxiliaries Department under Pobjoy, produced a series of sleeve valve engines designated PMS/1, PMS/2 and P6. ('P' indicated Pobjoy, 'M' was for Military and the numeral, the number of cylinders). Exactly when the decision to press ahead with these products was made is not known but it would not have been taken lightly. The Department had been created mainly to rationalise the problems of gearboxes and Rotol's expansion into the realm of the engine manufacturers was probably viewed with deep suspicion by the Directors of Rolls-Royce and Bristol Aeroplane Company and was only to be tolerated during the wartime emergency. Some day in the future, there would have to be a reckoning, but for the moment the Directors appeared to have given Pobjoy his head.

Although there was an early four cylinder version, the first engine of note in the series was the P6 Auxiliary Generating Plant, more commonly known as the AGP P6. It was designed to provide AC and DC current for aircraft services; a six-cylinder horizontally opposed ('Flat six') engine, rated at 60 bhp directly coupled to an alternator, and a train of gears drove a DC generator. Altogether it was a most creditable effort from a small team but when the time came for it to be assembled for its initial testing, someone with sufficient background knowledge to appreciate the finer points of sleeve valve engines, noted that the ports on the sleeve appeared to be machined incorrectly. Percy Pugh, who worked in the Experimental Shop and was a most experienced pre-war engine builder, realised at an early stage of the build that all was not as it should be when the ports in the sleeves persisted on opening when they should have been closing. Roger Armstrong, who was working in the auxiliaries department at the time, kindly explained to the author, with the aid of some rough diagrams, how the basic fault lay in the bevel gears which were driving the train of gears, and how this had the effect of reversing the drive to the sleeves.

'Panic stations' followed this discovery, as the date for the test, to be witnessed by the Ministry, was not far off. By a huge effort from all concerned – principally from the Experimental Machine Shop – the mistake was rectified and the tests successfully carried out. At the subsequent inquiry it was confirmed that a basic error had occurred in the original design, and there is a suggestion that the person responsible was demoted. Mistakes of this nature are unfortunate and happen on rare occasions, particularly when under pressure.

The story of the sleeve valve engine, reputed to operate in the reverse direction from the original design, continued to be related for years afterwards. At the time of the incident, some comedian in the Experimental Shop made a 30″ diameter medallion and hung it from the balcony rail, inscribed with the words 'To D.R. Pobjoy – for services rendered'.

The AGP P6 was unveiled at a gathering of the press and air experts at a luncheon held in the Savoy Hotel in London. (This was the same occasion when the change of company name was announced as reported in the previous chapter.) It attracted a great deal of publicity, not only in this country but all over the world and there was much speculation as to whether it would be developed for the post-war motor car industry. Although Rotol was not mentioned due to wartime restrictions, it was also the subject of an announcement on the BBC six o'clock news on Thursday 28th October 1943 – the date of the luncheon.

The unit was designed primarily for the Short Shetland flying boats, which were built to provide long-range protection for the Atlantic convoys, but only two aircraft were built by the time the war drew to its close. Pobjoy had put forward the idea of fitting the AGP P6 in the Shetland, these aircraft being large enough to accommodate two of the Rotol units; each weighed some 550lb and the soundproof box which encased it was a little under four feet in length. Unfortunately, on 28th January 1946, someone in the RAF duty crew decided to cook breakfast and started the AGP unit without checking to ensure that the ducts, which provided essential air-cooling, were in the open position. Consequently, the unit overheated and started a fire, rapidly spreading to the rest of the aircraft. It sank at its moorings at Felixstowe, and this seems to have been the end of the road for both the Shetland and the AGP P6.

The other two auxiliary generator units were much smaller affairs, intended as portable power units. The PMS/1 was a 4 hp single-cylinder unit providing power for battery charging, weighing around 75lb while the PMS/2 was a twin-cylinder 8 KVA Petrol-Electric Generating Set of much larger proportions. It was in two distinct parts, each weighing a shade under 150lb and was specifically designed for use in the jungles of the Far East, but it could also operate in sub-arctic conditions. Provision was made for it to be transported by four native bearers. The engine was in one unit and a light alloy gearbox driving three generators in the other, each within a tubular guard frame. A production order for 50 of these machines was completed but nothing is recorded of the /1 version. It is thought that a small pre-production batch may have been produced but work ceased on both these projects, when the war against Japan ended. The small number already built were probably sold off as war surplus, as the following story shows. In 1959, Vernon Christie was driving back home, after a business visit in the London area, when he gave a lift to a tall, athletic young man, carrying a large rucksack. He was a trawler fisherman from Iceland and while his boat was having a refit in Hull he was spending a few days hitch-hiking around the country. During the course of their conversation, as they neared Gloucester, he mentioned two generator sets which he had on his boat, one operating the radar set, the other a spare, with the name Rotol Gloucester on a brass plate. From his description it was obvious to Vernon that it was one of the PMS/1 sets which he had helped to design more than ten years ago. The young man was almost as amazed as Vernon to hear of this coincidence, and was full of praise for such a fine piece of engineering.

We come now to one of the lesser known products with which Rotol was associated. As we have seen, the 'improved' wood used for the propeller blades was made from a compressed laminated material, obtained by heat and pressure exerted on alternating layers of wood and glue. For best results, the temperature throughout the block of 'improved' wood needed to be as uniform as possible, but this was impossible with normal conduction heating, where the interior was always much cooler than the outside surface. In the method that was to supersede conduction heating, the reverse was true, a fact not readily appreciated by some of the people who were to use it.

Radio-frequency heating was known for some while before Rotol began to make use of its properties in the latter years of the war. It possessed several advantages, including a much reduced heating time, illustrated by the fact that it took twenty hours to heat a block of wood by normal methods but only thirty minutes using the RF system. The Research Department, which covered a wide field of enquiry, iniated the RF heating and attracted several personnel from other parts. Don Bishop transferred from the Main DO and went to work in the Research Department in 1943, where much of his time was spent in developing the glued blocks from which the wood blades were made, using the new RF heating. After a few months, Don moved to Hordern-Richmond with this technique, where he assisted in the development process for the wood blocks. He was fortunate to be among the favoured few who were allowed precious petrol (at 1/11d. per gallon or almost 10p) for his car and could make the journey back to Gloucester at the end of the week. On one occasion he was offered a lift in the Hordern-Richmond aircraft, a small twin-engined three seat monoplane known as the Autoplane, which deposited him at Staverton in record time.

It was realised at an early stage that one of the applications of RF heating was in the process of cooking and those employed in the Rotol Development Department at the time conducted some unofficial experiments, not all of which were considered entirely successful. John Cripps related a story of the early days of RF heating, when the canteen lady came round with tea and plain buns. Someone suggested they might cheer them up by putting them in the oven, so this they did and the results were much enjoyed. The next day, somebody came in with a meat pie and wanted to heat this up too. No one knew at this stage that RF heating does not brown the pastry and as they looked through the glass at the pie they saw its outward appearance was unchanged. The controls were slowly turned up and the outside began to glow as it became red-hot! Needless to say, the meat interior was completely crisp – done to a cinder in fact – which didn't please the chap whose pie it was.

Reg Streete was in charge of the Electrics Laboratory and was responsible for the RF heating development by the company. At the outset everyone seemed very pleased with the new product and there may even have been tentative plans for marketing a 1kw 'radio frequency pre-heater' in the post-war period. It was first mentioned at the Directors' Meeting of September 1943, when it was noted that 'if successful, not only would this produce a more uniform board, but eventually it would enable boards to be pressed to size'. The boards to which they referred were of course the wood blade blocks. As we shall see in the next chapter, RF heating at Rotol was not to survive into the post-war period.

This is an opportune moment to bring the Research and Development Department into the narrative. At the outset it comprised a number of different sections, all under the control of E.G. Fairhurst, the brother of LGF Both had been Rolls-Royce employees, but unlike his brother, EGF did not make the move to Staverton until June 2nd 1939. His initial duty was to continue to keep an eye on the Rolls-Royce interest at the new factory, reporting at a weekly meeting at Derby on the progress being made on the testing of the airscrews for the Merlin engines. Later his duties were extended to cater for other engine manufacturers, including Bristol Aeroplane Company, Armstrong Siddeley, Alvis and Napier, all of whom were to fit Rotol propellers to their engines in time.

The Research Department Engineer, responsible for the design of the test rigs and test hangar, in conjunction with the Plant Engineer was R.M. (Dick) Parkinson and his assistant was Jim Boyd. In charge of the various sections were A.G. Mapp (Instrument Section), Ken Boydell (vibration), R. Wimberley (hydraulics) and Messrs Streete and Tait (electrics).

A.G. (Alf) Mapp started in the Research Department in December 1940 with two fitters, and at the outset they occupied a small room known as the Instrument Room but this was enlarged within a few months and again in 1942, due to the ever-increasing volume of work. The design work was carried out by Alf Mapp and Norman Ramsey, with three technical assistants, Wagner, Evans and Alf Piercey. There were a number of machinists, among them Charlie Davey who had joined Rotol in 1942 as a toolmaker but transferred to the Flight Test Department in 1944 to undertake the fitting of prototype equipment, mainly on Rolls-Royce engines of various marks. The Section Charge Hand in the Research workshop was Len Preece, who later went over to the Marine Screw Department when they began operations.

Most of the early engine running was carried out in No 1 Test Hangar but not long after the war started two more were built in order to cater for the greatly increased volume of propeller testing. Bill Roberts was Research Foreman, responsible to Parkinson for the everyday running of the hangars and his right-hand man was Dick ('Wilkie') Wilkinson. 'Wilkie' had been chief mechanic for the famous bandleader, Billy Cotton, and liked nothing better than to relate stories of pre-war racing incidents. He held several track records and resumed his love of motor racing when the war ended, joining Jackie Stewart in Scotland. The three Rotol test hangars came under the overall jurisdiction of Eric Fairhurst and were in almost constant use during the war years. On one occasion he received a visit from the local police officer, following a complaint from a nearby resident that the noise from the test hangars was such that she was unable to stand it a moment longer. Eric pointed out that the work on which the company was engaged was vital to the war effort and he would only comply with an order from his Chief Engineer – the position occupied at the time by his brother, although this fact was not divulged. The policeman went away, but one can sympathise with the lady concerned to some extent as the continual roar of the engines and the high-pitched whine of the propellers, both day and night for hours and indeed days on end, must have made life difficult for anyone living nearby. There were few occasions when the hangars were not in use, as all the propellers received full development testing, followed by a further and lengthy MAP Type Test, in order to receive official approval before production could start.

At the beginning of the war, Alf Mapp and his small band were responsible for providing the necessary measuring equipment for the Research Department and others in the factory, probably the most important being the provision of Strain Gauge equipment to measure the vibrations of aircraft propellers. These were cemented to the surface of the blade and connected electrically to the measuring equipment, the design of much of this being in the hands of Tony Evans. In 1943, the section used a Morris 10 cwt van where the two operators would sit side by side at the control panel located in the rear of the van. The van, which continued to be used for a great many years, was particularly useful in that it was possible to measure the stresses under actual running conditions, with the propeller either mounted on its appropriate engine in the Test Hangar, or on an aircraft picketed down on the ground.

In 1941 the Electrical Laboratory was housed in a small area of the Experimental Shop at the foot of the stairs. It was surrounded by a meshweld and boarded partition and had a small screened room in one corner, which was much envied by some other departments, according to Arthur Hayward, who had just joined the company. A list of personnel at this time included Gordon Tait, Len Frost, Ron Marriott, Albert Morgan, Reg Alexander, Bill Evans and 'Pluto' Hayes. At this early stage, most of their work consisted of testing the hub motor of the Rotol Electric Propeller. Reg Streete ran the Electrical Laboratory from one of the offices on the balcony, some of which were also used by the Research staff and others by the Experimental Manager, Norman Payne and his staff. It was fortunate that all these different projects and individuals were able to work together so amicably, as it was quite a cross-section of the Rotol workforce.

The Electrical Production Benches, commonly known as 'the rails' were located in the Experimental Shop, under the supervision of Jack Jones, with Lew Richards as foreman and Reg Beard as leading hand. Assembling electric motors, small harnesses and suppressor boxes, was the job of a small workforce, largely female. Among the girls working there was Rene Griffin, later to be remembered for spending her latter years at the company in the Drawing Stores of the Main DO where she had a fantastic memory for individual drawing numbers. She remembers working alongside a number of girls from Blackwood Moreton, making up electric harnesses, who were being trained at Rotol before going back to get on with the contract in Scotland. Doreen Long was the senior girl and she and Reg Alexander were married in 1946.

When Arthur Hayward joined Streete in the Electrical Laboratory, there was a move to the other side of the Experimental Shop where an upper floor was constructed to house the General Office, with the working area and the screened room below. All this disappeared at the end of the war and no trace now remains. Over the next twelve months the Research Department expanded rapidly, many new faces appearing on the scene, a number of new projects were started, among them was the idea of RF heating, which started off this section of the history.

The question of using the variable pitch propeller to operate in reverse pitch and act as a landing brake was still in its infancy during the war, and although still not widely used, some aircraft of the present day have this feature. Rotol fitted a Halifax Mk V with four reversing propellers for development purposes in the latter years of the war and much of the testing was carried out on the grass airfield at Baginton. Ground runs were done initially at Staverton and it was during one of these trials that Eric Fairhurst witnessed the Halifax taxiing slowly down the perimeter track, followed by a cyclist some distance behind. The aircraft stopped, revved its engines and began to reverse back in the direction from which it had just come. The cyclist, completely unprepared for such a strange manoeuvre, fell off his machine and hastily scampered to the safety of the nearest hedge.

The ability to 'drive' the Halifax in reverse was sometimes used to impress visitors, as a kind of 'party piece', according to Bryan Greensted. The aircraft would be parked, nose-on, as close to the hangar doors as possible, with the tailwheel swivelled through 180° and the wheels aligned with the short taxiway that extended from the hangar to the perimeter track. At the end of their visit, the party was safely positioned on the tarmac outside the hangar, the Halifax's engines started and the aircraft was slowly taxied backwards along the taxiway, to the great astonishment of the onlookers. The fitters usually watched this performance with keen interest, to see how many of the VIP's would have their hats blown off!

Another incident concerning a Halifax was recalled by Charlie Harmer of the Flight Test Department. A visiting Halifax overshot the runway and ended up with its nose halfway over the main Cheltenham to Gloucester road. Double decker buses and other traffic had to be carefully driven around the obstruction and although it was a Sunday morning, a considerable amount of traffic was held up. The Flight Test staff leapt into action and attached their refuelling tanker to one main wheel, an RAF fire engine to the other and the Flight Test tractor to the tail. All three vehicles quickly brought the aircraft out of the ditch, in fact a little too quickly for the fire engine which was towing in reverse and couldn't keep up with the pace of events and the fire engine's crew hurriedly abandoned their vehicle as the tail of the Halifax bore swiftly down upon them. In the event there was only slight damage to both the fire engine and aircraft from the resulting collision.

Rotol fitted a reversing propeller to a Hurricane for experimental purposes, and the aircraft was flown to an RAF station, where the pilot landed and parked in a rather restricted space. When he returned, a small crowd of airmen were waiting to see his departure, no doubt wondering how the pilot was going to remove the aircraft without the use of a tractor. They watched with open mouths as the pilot got in, revved the engine, reversed the propeller and slowly backed the Hurricane out of its parking slot and took off.

Mastering the technique required to operate an aircraft fitted with reversing propellers took some time, as it needed careful co-ordination of pilot and engineer. On touchdown, the pilot would call, 'Reverse' and the Flight Engineer, with a single movement, would pass all four propeller pitch control levers into the reverse position. At the same time the throttles were gradually opened to full power, taking care that no uncontrollable swing developed. With reverse thrust being produced, any swing was difficult to control, as any reflex and immediate correction by use of the throttles might only accentuate the swing. Trials with reversing propellers continued for some years with varying degrees of success but following a bad crash at Moreton Valence in 1950, which will be related later, Rotol seems to have given up further development.

After almost six long years, the hard slog of life in wartime Britain came to an end in 1945, with victory in Europe on 8th May and the final surrender of the Japanese in August. It was a time of universal rejoicing for many, with the welcome return of men and women from the armed forces, especially those who had been prisoners of war.

Rotol are reputed to have produced some 100,000 airscrews by the end of the war, and they had flown on over sixty types of production aircraft – on more than a hundred if the prototypes are also included. But there were doubts about the future of an industry which had depended largely on Government contracts for its survival. It was obvious that Rotol could not sustain the number of workers which had prevailed during the war, and reductions were inevitable. The exact size of the redundancy that subsequently took place is not known, although a figure of 15% has been quoted. This is unlikely to have included the many Rotol dispersal offices and factories up and down the country, which were closed down fairly quickly following the end of the war.

The Repair Department at the Gloucester Wagon Works returned to Main Works on 6th May 1945, and the Auxiliaries Design Office at Imperial House in Cheltenham and the Technical College in Gloucester, were handed back to their owners soon after, and all Rotol staff came back to the premises at Main Works. The factories at Ossett and Worcester, the Spinner Factory in Cheltenham and the Auxiliaries Repair Department at Roberts toy factory in Gloucester, all closed down a few weeks later. If the machines were Rotol property, they were returned to Staverton, but it is assumed that the employees who had been recruited locally for the outlying factories must have been made redundant. Company employees who had gone to these far-flung outposts from Main Works, on the understanding that they would be offered their jobs back when the war was over, returned to Main Works but this in itself merely exacerbated the growing problem of over-staffing. Coverley reported to the Directors that more than 600 employees would have to be reinstated following closure of the dispersal centres.

On their return, they were sometimes offered a different job from the one they had been doing during the war, with a lower salary, which is what happened in Ted Marshall's case. He had been in a responsible position at Ossett and the job he came back to involved a down-grading, which he refused. Instead he went to Dowty Equipment and some years later, after the Dowty take-over occurred in 1959, found himself back at Rotol, this time in the Estimating Department.

Frank Morris, who had been in the Spinner Factory in Cheltenham, found himself in a somewhat different position. When he came back to Main Works, there was so little to do that he occupied his time making toolboxes for the men and selling them at 25/- (£1.25) each. Eventually he managed to get a transfer to the wood blade shop, but the days of the wooden blades were clearly numbered and Frank ended up in the Maintenance Department, where he recalled tackling some interesting jobs in the houses of the directors before his retirement in 1977.

A certain amount of re-shuffling of personnel followed, and every effort was made to find alternative work for many of the employees in other parts of the factory and offices, where happily, some of them went on to finish their working lives as has been testified during the research for this book. It is assumed that a large number of women workers were quite pleased with the redundancy, which enabled them to return to the domesticity which had been their lot before the Ministry of Labour had decided otherwise.

It had been a condition in the conscription act, that men and women joining the armed services would recommence work at their original place of employment on returning to civilian life. This was the theory but in practice it might result in a different position being offered than the one that had been left. Jack Dancey, who had been among the first employees in 1937 at Llanthony Road, was demobilised in 1946. When he left to join the army, he had a staff appointment in the Accounts Department, so he was not too pleased to be offered a job in the works on his return. He stuck it for about a year but during this time he was conscious of the over-staffing resulting partly from the return of those employees from outlying factories and the Rotol men and women coming back after wartime service. Coupled with the reduction in propeller contracts, there was insufficient work for them all, and he often used to walk aimlessly around the factory with a piece of paper in his hand, meeting others on the same mission. After a while, he could stand it no longer and gave in his notice. One also should not overlook the unsettling nature often experienced by the men and women returning after serving for up to six years in the armed forces, many of whom found great difficulty in accustoming themselves to 'civvy street' and its so-called freedom. With the war over and demobilisation now in full swing, some may be surprised to learn that conscription continued for a time, to be soon replaced by National Service for a fixed period of two years. Stan Price, one of the first Rotol apprentices, received his call-up papers for the RAF in mid-1945, and he and about thirty other young men from Rotol met on Gloucester station when they travelled to the recruitment centre near Manchester. They had all been exempt from military service during the war, because of the nature of their employment. A new situation now prevailed and they were needed to take the place of the many thousands who were coming back to resume civilian life.

1945 – Dramatic Changes at Rotol

The year 1945 was to see a great upheaval in the work force, not only in the lower ranks but also in the top tier of management, as several well-known figures severed their connection with the company. Although she played only a relatively minor role in the company, Miss D. Ashworth who left in the middle of the year will be remembered by many with some affection as she had been the Women's Welfare and Employment Officer at Rotol since 1942, when she had taken over from Miss Evans, who had a reputation for being something of a martinet. Before the war, Miss Ashworth had assisted in the organization of International Camps for Girl Guides and had travelled extensively in both the Middle East and the Far East. *The Airscrew* described her as 'always showing a sympathetic understanding towards those who came to her for advice and counsel'. Miss M. Harris, her replacement, had been her understudy for the previous two years.

Some weeks before the European war ended, the General Manager, Mr R. Stammers, departed. At the Board of Directors' meeting held on March 15th 1945, the minutes recorded that, 'due to certain circumstances . . . the details of which are known to the directors, it was agreed to dispense with Mr Stammers' services. In recognition of his continued employment with the company as General Manager since 1937 it was decided to pay him £3,500 as compensation for loss of office.'

It may be that Stammers breathed a sigh of relief as he left. His had been a particularly arduous task at Rotol as he always stood in Coverley's shadow, rarely able to exercise a free hand. He was certainly not the human dynamo figure portrayed so convincingly by Coverley, which probably explains why he had been pushed aside by the Directors in 1939. From discussion with people who knew him, the picture which emerges is of a man noted for his quiet personality and good manners, but these qualities may not have 'cut much ice' with his compatriots at Rotol. He is believed to have become General Manager at Lagonda Motors, and later joined the Board of Fielding and Platt, a local Gloucester company, where he died quite suddenly in the 1960's.

Robert Henry Coverley, the Rotol Controller for the whole period of the war, was the last of the senior management to leave. Here was a completely different character from Stammers, or indeed anyone else at Rotol for that matter and we need to look back some months to obtain a clear picture of the circumstances which made it inevitable that RHC would sever his connection with Rotol.

For some time prior to 1945, discussion on the future policy of the company had taken place at the Directors' Meetings. Although there was a considerable reduction in propeller equipment for the Ministry of Aircraft Production, due in part to the huge influx of American aircraft, other products were still being developed. Some of these have already been related, including auxiliary power units and cabin air conditioning and finally radio frequency heating, for which a large demand was anticipated. Action had already been taken to adjust manufacturing capacity, by the withdrawal of work from sub-contractors (their numbers had been cut from 400 to just 34) and in addition, working hours in the factory had been reduced from 51½ to 47 per week. At the same time, with the closing of the dispersal premises in Cheltenham and Gloucester, the company were having difficulty in finding sufficient space to accommodate the returning departments. The high interest shown in the electrical products necessitated an enlarged Research and Development Department, but where to establish this was a problem that could not easily be solved.

The answer to this – and most of the other problems that Rotol faced when the war was won – was given by the Directors in a letter they sent to Coverley, dated 5th June 1945. This most important document was the main item for discussion at the meeting which followed on the 14th June. Although no copy of the letter survives, we can be reasonably sure of the gist of its contents, from the minutes of the meeting.

Much of the time at the meeting was spent in a somewhat one-sided battle between the four Directors, White, Sidgreaves, Hives and Rowbotham, on the one hand and RHC on the other, endeavouring to justify the wartime actions taken by the company, in the face of some heavy criticism. Although Cameron was also present, the minutes do not record that he was asked to contribute to the very frank discussion which took place.

Sidgreaves, the chairman, read from a prepared document, outlining the original reasons for the formation of the Rotol company and stating that the company's policy must be the production of the best propellers, both for aircraft and marine work. The report continued, by stating that Rotol had technically slipped behind its competitors in three ways: by the continued use of wooden blades, by the length of time taken to clear the contra-rotating and the reversing propellers, and by faults in present propellers, such as imperfect balancing.

Coverley stoutly refuted all these charges, reminding the directors that wooden blades had been the saviour of the country. During the Battle of Britain, magnesium blades had proved to be unreliable and the shortage of duralumin had given Rotol little alternative but to develop the wood blades. Furthermore, they had proved to have a great advantage over metal blades, due to their easy repairability and the fact they would not cause further damage to either the engine or the aircraft, should they be seriously damaged themselves.

The discussion, which was probably fairly heated, continued with the Directors stating quite firmly that they would not invest further money in the manufacture of wooden blades. They were also not prepared to invest capital in additional plant for either electrical equipment, small auxiliary generating plants, or the production of gearboxes. Both parent companies would ultimately require gearbox production in their own workshops, a point they had made on many previous occasions.

At the conclusion, Coverley was requested to prepare a review of the expected output in propellers, together with the projected requirements for plant and machinery in the next few months, to put before the next meeting.

Coverley's report on future policy, dated 23rd July 1945, was considered at the August meeting of the Board of Directors, and although no copy of this report has survived it was obviously a very comprehensive document. It had been circulated to the Directors prior to the meeting and they were fully aware of its contents. The Chairman fired the first shot by reiterating the earlier statement, 'that the parent companies were only willing to consider providing further money for the development and production of air and marine propellers.'

In reply, Coverley said that if such a decision was carried out, it meant the virtual closing of the factory. His estimate for the future aircraft propeller programme was only 25 per week, which was obviously insufficient to maintain the factory in its present form. It had already been agreed that the big market for marine propellers (if indeed there was such a market) was still some years away, so no salvation lay in that direction. He argued that the new business of RF heating should be vigorously pursued, as the company had received 196 enquiries for RF sets, in addition to the possible order for 50,000 motors for refrigerators that was waiting for a policy decision.

In one section of his report, Coverley argued for the parent companies to be prepared to provide further capital up to a figure of £1,350,000, in order to finance not a cut-back but an expansion programme. This was fairly obviously not going to find favour with the Directors, and from the general tone of the discussions that followed, Coverley must have realised it stood no chance of acceptance. Many of the arguments which had taken place at the previous meeting were aired once more and finally a decision was taken, that if Rotol could find ways and means of providing the necessary capital – without assistance from the parent companies – to finance the other items (the RF heating, auxiliary generators etc), then there might be a possibility of the Board changing its opinion. Coverley must have known that the chances of achieving this miracle was highly unlikely, but he took the suggestion away, to be mulled over by the Rotol management.

The last Board Meeting attended by Coverley took place on 11th October 1945. A comparative statement of turnover was tabled for the nine months up to September 1944 and the similar period for 1945. This showed a figure of £10,153,635 for 1944 and £4,975,734 for 1945 – a reduction of £5,177,901. Cameron reported that the overhead rate was rapidly increasing owing to the decrease in output and that the MAP had agreed to an omnibus redundancy contract. (Presumably this was some form of financial assistance, with severance pay, to the workers made redundant.) All this was depressing news but possibly only what one might have expected, given the known circumstances.

The first part of the Controller's Report, also included another unpalatable fact, that actual work remaining in the factory was much less than indicated on the programme, due mainly to the substantial number of components being finished ahead of schedule.

The second part covered future policy, and the chairman, Sidgreaves, began his remarks with a warning that Coverley should not take any of what was to follow as a criticism of his efforts and the results which had been obtained during the war period. (The reader may be forgiven at this point if he or she feels a little cynicsm on reading these fateful words, which seem to have a familiar ring.) These had been excellent, and something for which Coverley and the Company could be proud. But he stressed once again the view of the Directors, which remained unchanged, that the company had dissipated its efforts on miscellaneous items, thereby slipping back on its design, development and manufacture of propellers. He stated that the past was finished, and from this date the company would cease work on all items other than air and marine propellers, and that this view was endorsed by the Ministry of Aircraft Production.

The future of the Hordern-Richmond factory, which was still producing wooden blades, was also discussed. The Directors had decided that all their work should gradually be withdrawn and any uncompleted orders brought back to Rotol, with a target date for final completion, of March 1946. It had been agreed with Rolls Royce, BAC and also Supermarine, the makers of the Spitfire, that only metal blades would be required in the future.

The meeting dragged on, with Coverley attempting to convince those who sat opposite him of the soundness of his argument. He complained that the decision to restrict the company's activities would severely reduce his remuneration (his salary was partly based on the profit shown by the company at the end of each financial year) but he was told that everyone had to face up to a salary reduction in the present climate. Coverley also pointed out that he had been promised the post of Managing Director but the chairman denied any knowldege of the fact.

Sidgreave at one point stated that Rotol should have had the best propeller at the end of the war, instead of which they had finished with the Rotol design well behind its competitors. This was manifestly both unfair and completely untrue, and perhaps was only said in the heat of the moment. No doubt Sidgreave regretted it almost at once. One can imagine that by this time several similar remarks were being made, as claims and counter-claims were flung across the table. It is unfortunate that this kind of discussion almost always degenerates into 'a slanging match', no matter how carefully one chooses ones words at the outset, and this was no exception.

In order to clarify the position regarding the statement that the Rotol propeller was inferior to its competitors, it should be noted that the Rotol hub had a large shank diameter to accommodate the wooden blade, and the shank had to be somewhat longer, because of the ball-bearing stack used in the original design. This was a legacy of the collaboration between the Bristol Aeroplane Company and Curtiss-Wright, handed over at the inception of Rotol. Both these features contributed to the Rotol hub being heavier than that produced by De Havilland.

Coverley replied by pointing out that although still in the early stages, the Design Department had already begun to re-design the hub in order to combat this deficiency, incorporating Timken taper roller bearings for blade retention. It had first been brought to the attention of the Directors at a meeting in May 1945, and he took the opportunity of reminding them of this fact.

The minutes of this meeting cover ten foolscap pages, with a surprisingly detailed account of what was said. Although much of it is interesting to the historian, the ordinary reader will soon appreciate that Coverley was merely fighting a rearguard action. He was never in a position to alter the policy which the four directors had adopted, but it was not in his nature to give in without a fight and he did the best he could in the difficult circumstances. The meeting closed with another reminder from the Chairman that they were dealing with the future and not the past, and Coverley was asked if he fully appreciated the position. He was secretly of the opinion that the company had no future if it was to be restricted as outlined by the Directors, but he left the meeting without making further comment.

There was really only one outcome, as Coverley must have realised. Completely at odds with the board of directors, he had no option but to resign. When he requested that some of his personal office furniture be sold to him, the directors curtly dismissed the idea. He received a gratuity of £7,500 for his services, which terminated on 15th December 1945. Thus ended a period at Rotol, the like of which was not to be seen again.

What happened to him in the succeeding years is unknown although there is some speculation that he started his own company, and is believed to have died in 1954. Although possibly disappointed with the way his career ended at Rotol, Coverley must have gained some satisfaction from the award of the CBE that he received in the 1945 New Years Honours List.

He asked Miss Jackson, his personal secretary for all the years he was with Rotol, to accompany him, but she declined. There were many who wondered how she had managed to put up with him for so long, but Miss Jackson loyally insists there was a more human side to his nature that few could see. Dennis Savory supports this view – on occasions he was directed by RHC to hand deliver some important mail to the Montpellier Post Office in Cheltenham, and was rewarded with half-a-crown on his return. Miss Jackson stayed at Rotol for another eighteen months as personal secretary to John Reid, who took the appointment as Rotol General Manager (Stammer's old position) and she left in 1946 to read English and Philosophy at Bristol University. It is interesting to note that when RHC left Rotol the title of Controller also disappeared.

Because of the directors' decision to close down both the auxiliaries department and a part of the electrical department, notice was given to two senior staff members, D.R. Pobjoy and R.H. Streete. Pobjoy left on 30th November with a gratuity of £2,500 and Streete received £1,000, which denoted their respective value to the company. Frost, Goodfellow, Tait and Whitticase, all members of the electrical department, were dismissed with Streete, each receiving a variable gratuity, a little over £150. All these personnel had service agreements with Rotol, which contained a financial clause covering this event. The majority of the men dismissed from the electrical department were associated with RF heating and most of them moved to Grantham, where they started a firm to continue the process they had developed at Rotol. Electrical personnel engaged on synchroniser work and similar products associated with the aircraft propeller side of the business stayed at Rotol.

Prior to the resignation of Coverley, there occurred an unexplained and somewhat curious incident. A trio from Rotol, comprising Messrs Fairhurst, Danvers and Bubb were summoned to a meeting with Hives at Derby, not knowing what lay in store. When they met, Hives levelled the criticsm that Rotol had strayed too far from its propeller origins and that henceforth the auxiliary gearbox work would be done by the engine companies, with Rotol concentrating solely on propeller contracts in future. All three of the Rotol executives were startled to hear this pronouncement. They were under the impression that because the company had enjoyed a great deal of success with auxiliary gearboxes, further good times lay ahead in this field. The parent companies had probably also realised this and were merely ensuring that their own workforces would participate in any future successes.

The three men from Rotol departed from Hives office in a somewhat subdued frame of mind. To cheer up their flagging spirits on the return trip, they stopped off at the Queen's Hotel in Kenilworth, to have, in LGF's words, 'a right royal do'. When the bill arrived, LGF found to his dismay, that he had insufficient funds to meet it. He appealed to the others, who, despite being somewhat short of cash themselves, just managed to cover it. After they had departed, George Bubb confessed that he had left the only remaining money he had as a tip – it was just three pence! Although he subsequently passed it many times, LGF admitted that thereafter he never had the nerve to set foot in the establishment.

The meeting at Derby with Hives and the three men from Rotol raises an interesting point – why didn't Coverley tell his Chief Engineer, L.G. Fairhurst, what had transpired at the meeting with the Directors? Whatever the explanation, it is certainly difficult to understand Coverley's motive in letting them make the trip to Derby with no knowledge of the reception they would receive. From correspondence with LGF it is clear that there was little love lost between these two major Rotol figures. The reason for this is almost certainly because Coverley could not ride 'rough shod' over LGF as he did with so many others and indeed made several attempts to remove him from office from time to time. Fairhurst had a philosophy throughout his engineering career whenever he was in a difficult spot of 'standing his ground, like a Daniel' from a Salvation Army Hymn.

Dare to be a Daniel,
Dare to stand alone,
Dare to have a purpose firm,
And dare to make it known.

There were few men in power at Rotol who were prepared to adopt this stance with RHC.

With the winding-up of the Auxiliaries Design Department all the staff came under L.G. Fairhurst's control, who took the opportunity to reshuffle some of the personnel. The draughtsmen and designers were easily accommodated into the existing organisation, as were the other technicians. George Bubb became Chief Development Designer with Mrs Robjohn as his secretary. Vernon Christie took on the newly created position of Installation Engineer, with a small team of technicians to cover this rather complex subject – fitting the proverbial quart into a pint pot has always been one of the main headaches in designing parts for aeroplanes!

One direct result of this decision by the Directors was the handing over in the half-finished state of the Bristol Brabazon gearbox that Rotol had been working on for some time to another manufacturer – possibly the Bristol Aeroplane Company themselves. However, according to L.G. Fairhurst, because of pressure from the Ministry, the company were persuaded to reinstate the design, development and manufacture of auxiliary gearboxes. The date of this reversal of policy has not been established with any degree of certainty, but it cannot have been long after the original Board decision, as there does not seem to have been a break in gearbox production. (The minutes of the Board of Directors' Meetings covering this period are missing at the time of writing but if they ever come to light this point may be clarified.)

A new Auxiliary Gearbox Design Section was created, to join the other sections in the Main Drawing Office, with a new section leader, Eric Langton. The company went on to produce thousands of Rotol gearboxes in subsequent years for a host of aircraft, most of them individually tailored to suit each application. Because the installation of each box became so complex, the luxury of providing a universal box was never again achieved, nor was it possible to use many common parts, except for small items.

Passenger and transport aircraft were in great demand in the years following the war and Rotol made gearboxes for almost every British civil airliner of the period, including the Handley Page Hermes, Avro Tudor, Bristol Wayfarer and De Havilland Comet. The military side was also well represented, one of the largest orders being for the Gloster Meteor. This aircraft, in its many variants, was sold to air forces all over the world. Rotol gearboxes were also fitted to the Fairey Firefly, Avro Lincoln, Avro Shackleton, De Havilland Vampire, Hawker Hunter, English Electric Canberra and Gloster Javelin and this is by no means a complete list.

Strange to relate, thirty or so years after the Rotol Directors had stopped all further gearbox production, a similar decision was made, and the gearbox design team were disbanded. No doubt the reasons then appertaining were entirely different from those in 1945 but the result was much the same. After a short period, the company reversed their decision and gearbox production recommenced. At the time of writing, the company is actively pursuing a contract to supply an auxiliary gearbox for the next generation of European fighter aircraft.

A Nautical Excursion

With their expertise in aircraft propellers now firmly established, it was perhaps a logical step to turn to marine screws and this the company did in 1941. In common with all the new products which Rotol pioneered during the early war years, the first marine designs came from the Project Office over the Spinner Factory in Cheltenham, where, because it was always Rotol policy to draw the designs full size, a window had to be removed in order to get an extra large drawing board into place. Later on, with the larger diameter blades, the idea of drawing to a smaller scale was finally adopted, although it almost always caused some problems when the drawings got to the shopfloor.

The Marine Section first began operating from premises in the Technical College in Gloucester, moving to the newly constructed Technical Block when that was built. They finally ended up in a specially built section with offices alongside at South Works, in what later became No 6 hangar, where they took up residence in 1945. Manufacture – apart from the blades – and assembly were all done by the section personnel but blade manufacture was undertaken in the Production Shops. However, there is a suggestion that early blades were made at 'the toy factory' in Gloucester, where one of the Repair Departments operated.

In charge of the section was an ex-Rolls-Royce employee, Harry Twitchen. Harry had started at Rotol in May 1939, working in the drawing office under LGF until taking on the Marine Section. Some years later a well-known naval architect and marine engineer, Fred Cooper, joined the staff in 1943. He had been responsible for the design of many of the pre-war boats, including Sir Malcom Campbell's 'Bluebird', and at one time was Manager of the British Power Boat Company. The Rotol Installation Engineer, George Harvey, was responsible for all airscrew and marine screw trials and installation, and his senior engineers were Ernie Thomas and Bob 'Fearless' Foster, two of a number of car and motor cycle racing enthusiasts who came to Rotol during the war years. Both Ernie and Bob had both achieved fame in the Isle of Man TT motor cycle races before the war.

The harbour tug *Capable* was one of the first applications of Rotol marine variable pitch propellers and proved itself over a period of many years. The craft was fitted with a pair of three bladed propellers, 10′6″ in diameter weighing in the region of three tons, with a pitch range of 110°, from 90° ahead (in the fully feathered position) to 20° astern. On one occasion, the tug fouled some heavy mooring chains, damaging a propeller blade but after the propeller was stripped its operating mechanism was found to be in perfect working order. A particularly valuable feature was the ability to change a blade by removing six bolts, while the variable pitch mechanism remained sealed off. The craft could be manoeuvred within very fine limits which enabled the strain to be applied gradually when the tug was towing. A bollard pull of 27 tons was achieved with fixed pitch propellers but with the fitment of the Rotol installation this was increased to 36 tons.

Another notable success was that of the Fairmile Torpedo Boat, a D-class craft that first made its appearance in 1942. More than 30 firms produced various installations from the basic design for service with the Royal Navy, and by the RAF for Air-Sea Rescue work. It was powered by four 1,250 hp supercharged petrol engines, giving a top speed of 32 knots. Although Rotol only fitted variable pitch screws to a small number of the D-class boats, they gave a 50% reduction in fuel consumption and a 70% increase in bollard pull.

A wide range of marine craft, including tugs, motor torpedo boats, firefloats, submarines and air-sea rescue launches had Rotol marine screws. They were powered either by petrol, diesel, steam or gas turbine engines ranging from 15 to over 6,000 horse power. On the whole, the company had an excellent record from the trials of these boats, both in performance and propeller control and the work that Rotol began in the 1940's has now been incorporated in vessels all over the world, including nuclear powered Russian ice-breakers and the cross channel ferries.

Several advantages stemmed from the use of variable pitch marine screws, perhaps one of the most important being the phenomenal reduction in stopping time and distance, which was generally halved. Harry Twitchen recalled that the first time this was used during sea trials, most of the crockery in the galley was broken! The ability to maintain station using propeller control was another, which proved to be extremely useful. The river firefloat was able to dispense with its water pump auxiliary engine, at the same time increasing pumping capacity. By using the pitch control (which needed only minimum main engine power) the craft could hold station alongside a fire, with pumps being driven via a clutch, by the engine.

One of the less successful applications concerned the air-sea rescue launch called the *Celerity* – which means 'Swiftness' – although the name did not appear to be well-chosen and seems to have been dropped from company usage. The craft, a 105 ft launch, was designed and made by BAC although the hull had been designed by the Chief Designer of Fairmile Marine in Surrey and was powered by four Bristol Hercules air-cooled radial engines. From the outset, concern had been expressed by Rotol at the cramped installation of the engines, as these were positioned in the bowels of the craft, so it was no small wonder that cooling became quite a problem. Following initial trials, the Rotol design team provided cooling fans, mounted in front of and driven by the engines. Unfortunately, the power needed for each fan was itself in the order of 90 hp which reduced the power available for propulsion from the main engines to such a degree that the design speed of the boat was not achieved and the project was abandoned.

The *Celerity* was responsible for one of the most widely-told stories ever to come out of Rotol. What follows may be regarded as the authorised version, as it was related to me by the man concerned, L.G. Fairhurst.

> In the early days of the *Celerity* project we had difficulty in finding a supplier for the stainless steel propeller shaft. Each of the four shafts was some 30ft. in length, 4″ outside diameter with a 2″ hole down its length, delivering the high pressure oil

> supply from the engine-driven pump to the actuating mechanism in the propeller boss. Eventually, we found a supplier and after some time the shafts were delivered to Rotol and booked-in to the raw material stores, where bars were stored vertically in racks. I went down to the stores to see these monumental pieces of engineering and asked a workman if he knew their whereabouts. He pointed to one of the racks. 'There they are. They were too long to fit in the racks, so I had to saw them in half – my boss is always complaining about leaving stuff lying about on the floor'. You can imagine my feelings as I returned to the office, repeating to myself – he sawed them in half to fit the rack . . .

Reg Russell was working in the Raw Material Stores at the time of this incident, although he maintains that the sawyer, who may have been Johnny Boal, an ex-jockey, cut it into several short lengths. This is only a detail, whether it was in two or twenty two parts made little difference to the outcome.

LGF had to report this disaster at the next meeting with the Admiralty at Bath, a few days later, when the story was greeted with some incredulity. Fortunately, further supplies were promised in about three weeks time, so the project was not delayed unduly. One can assume that the stores were left in no doubt about the advisability of keeping the length intact, on all future deliveries of this particular item.

This story soon went all round the industry and beyond. LGF even heard it recounted by a US Lieutenant Commander, at a dinner in the naval base at San Diego, where it caused great amusement and it seems to be the one story that Rotol old hands relate whenever they get together.

In addition to those fitted to surface craft, Rotol also made four variable pitch marine screws for fitment to a Royal Navy submarine, HMS *Truculent*. Rig testing and initial surface trials had gone well and preparations were made for the submarine to make its first dive, using the new marine screws. Lloyd Jones had spent the war years as a service engineer for BAC and worked all over the world, mainly on the Hercules engines fitted to a large number of British aircraft. He joined the Marine Section in 1947 as a Service Engineer, and was looking forward to being with the craft on this dive, having been intimately connected with the installation from commencement.

On the day when the first dive was to be made, Lloyd found that his insurance did not cover him for this activity, and despite his annoyance, the submarine set off without him. On the way down the Thames, a collision occurred with a Swedish ship and the submarine was sunk, with a great loss of life. As his normal post would have been in the engine room it is doubtful if Lloyd would have survived. He telephoned the police in Prestbury, where he was then living, with instructions to tell his wife he was safe and well. Unfortunately, by the time the police arrived to deliver the welcome news, she had already heard of the sinking on the radio, and one may imagine her first reaction when she saw the policeman approaching her front door. Lloyd went on to serve a further 20 years with the company, and retired from the Inspection Department in 1973.

LGF had also advised Harry Twitchen to ascertain if his insurance covered him for underwater trials. As it didn't he was also fortunate in avoiding a rendezvous with the grim reaper. In 1951, Harry severed his connection with Rotol and joined Stone's of Deptford, who were marine engineers. It seems that

Rotol policy at this time was to concentrate on marine propellers below 4′6″ diameter and Harry was convinced this was a wrong decision. In his opinion, the market for VP Marine Screws lay in larger screws, but this would have meant the acquisition of much heavier machinery by Rotol, at a period when they were already short of funds for other developments. In time, Harry returned to the Gloucestershire area, where he took up employment with the Dowty Group and eventually retired in 1981. With his departure, Terry Elmes, who was already in charge of the Pentane Section, the details of which will be related shortly, took over.

Terry has a great fund of humourous stories, of which the following is an example. When the company began to take on university graduate apprentices, part of their vacation was spent in one of the technical offices at Rotol. On this occasion, Terry sent his graduate apprentice off on a week's course. On his return, Terry signed the lad's expense sheet but one of the items – 9d. for a pound of cherries – was queried by the cashier, who sent it to LGF for his comments. He returned the form with the terse comment 'A growing lad needs his fruit' scrawled across it, following which the expenses were paid in full.

George Ronson started at Rotol in February 1949 as a performance engineer, working exclusively on marine propellers, and was soon in charge of the special water tunnel, used for testing these products. One particular incident proved to George that there was quite a lot of work still required before the tunnel could be considered satisfactory. LGF was conducting a distinguished visitor on a tour of the plant and when they arrived at the tunnel he pointed out the armour-plated glass, enabling the test engineers to view the performance of the screws under water. To demonstrate the strength of the glass, LGF picked up a large hammer, and swung it violently against the glass – which promptly smashed into small pieces!

The most advanced variable pitch marine screw ever produced by Rotol was that used on the *Grey Goose*, reputed to be the boat on which Commander Scott (later Sir Peter Scott) served in the Royal Navy during the war. The *Grey Goose* had been modified to take two Rolls Royce 6,000 hp gas turbine engines, each engine driving a Rotol 5′6″ diameter vp screw. In 1955, HMS *Grey Goose*, at that time the only naval craft in the world propelled by gas turbines, took part in the British Trade Fair at Copenhagen. Keith Lloyd travelled with her as the Rotol technical representative and wrote of his experiences in the Rotol and British Messier magazine. *Grey Goose* was the sole British Naval exhibit and came in for a great deal of attention, from the Royal Danish Navy, the press and television. Demonstration cruises took place throughout the visit and during the two thousand mile journey the Rotol equipment operated with its usual smooth efficiency.

Most, if not all, of the marine screws were funded by the Admiralty, and although their Lordships appeared interested in the technical aspect of variable pitch marine screws, the idea was not accepted wholeheartedly. In the late 1950's it was finally recognised that there was little possibility of large scale orders and the Board called a halt to all further work. There is some suggestion that existing designs were sold to Stone's of Deptford, but this has not been confirmed. Whether or not this decision was sound with hindsight, the closure of the marine department .soon followed. The majority, if not all, of the personnel were absorbed into various departments within the main works. Terry Elmes had by this time taken on the post of Chief Designer from Tom Godden who had moved on in 1953 to become Chief Engineer. The Design and Drawing Office now had its hands full with the design of the propeller and gearbox for the Rolls-Royce Dart, as well as a variety of other projects, so the loss of marine work was accepted without complaint.

We must go back to 1948 to pick up the other nautical project which was to occupy the attentions of the company for a few years. Rotol were approached by the Admiralty Underwater Weapons Establishment to undertake the design, development and prototype manufacture of the 'Pentane' acoustic homing torpedo. (The reason for the choice of name is not known.) A special section was set-up on the South Works site, where the project was worked on in conditions of the utmost secrecy.

In charge was R.T. Elmes, at that time the Assistant Chief Designer, who had joined Rotol in 1938. Terry Elmes had the distinction of being the very first apprentice (Clock No 6) taken on by a youthful George Dowty in 1934, where he received the sum of 10/- (50p) per week. (The second apprentice was Robert Hunt, later to become Sir Robert, chairman of Dowty Group following Sir George Dowty's death.) After serving his time with Dowty, Terry went to Boulton Paul Aircraft for a short spell in 1938 before moving to Rotol, joining the DO Electrical Section under Charlie Palmer. When the Project Design Office was started a short time later, he was one of the founder members, where he gained the reputation of being the fastest designer in the office and consequently got more than his fair share of the interesting jobs.

The Pentane department comprised the design office with a dozen draughtsmen, a stressman (Norman Hall-Warren), a part-time performance engineer and a secretary plus a small workshop devoted to assèmbly and testing, in the charge of Jack Moore. The Experimental Shop at Main Works undertook manufacture of all the components. Although everything was under a cloak of secrecy, the existence of 'Pentane' was common knowledge throughout the factory, and Terry came to realise that some tightening of security procedures might be necessary after the following incident. Being always the last to leave, he made sure the place was securely locked up, but on this particular night he had to return for some reason. To his amazement he saw the building was ablaze with light, all the doors unlocked and open. There was no one to be found but he eventually located the forewoman cleaner who told him that when she came on duty her first action was to open up with the master keys, ready for the arrival of the cleaning ladies later that evening.

The Pentane torpedo was a remarkable feat of design, and a great credit to all who worked on it. Basically, it was an anti-submarine weapon, consisting of a war head, an aluminium hull, electric motor power unit and contra-rotating output shafts driving fixed pitch propellers. It had an all-up weight of some 2,000lb and a top speed of 30 knots. The hull was made up of a number of casing sections, each about two feet in length and the first few sets were machined from DTD683 hollow forgings. Nothing was known at that time of the tendency of this material to self-destruct under the influence of its own internal stresses. There was great consternation one day, when one of the sections, in a part machined stage, split open while lying on the machine shop floor. Subsequent hull casings were fabricated from 0.2″ thick aluminium sheet, rolled and edge-welded and stiffened with internal rings. Unfortunately, welding caused distortion, so Araldite was tried in its place. The first one fell apart, due largely to the failure to observe scrupulous cleanliness but after the assembly shop personnel were reminded to always wear the gloves with which they had been issued there was no further trouble.

The first six models were empty shells, filled with ballast to test out the strength of the weapon in drop tests. There were two test ranges available, one in deep water near the Isles of Scilly, the other in shallow water in the Solent. It was decided that the trials would take place at Gosport, where the torpedo was to be dropped into the sea from a Fairey Barracuda in order to check the angle of entry, which was one of the vital features of the Rotol calculations. On the first run, the torpedo failed to release due to some malfunction, and after some further futile attempts the pilot decided to return to base. On the way, over Southsea – at the height of the summer season – he tried again and this time the mechanism worked. It fell from 900 ft into five feet of water and ran onto the beach between the holiday makers, fortunately missing everyone.

At a subsequent trial, which Terry had the dubious pleasure of witnessing, they were on board an RAF launch, cruising up and down the Solent just outside the buoys which marked the dropping range. In charge of the operation that morning was a Squadron Leader who took pains to point out two things; firstly, his pilot was so skilful that the weapon would drop into the water absolutely 'bang-on' alongside the launch and secondly that no bad language should be used on the radio because it had been known for calls on their frequency to be heard on the BBC Home Service, which had resulted in some harsh comments from higher authority.

On the day of the trial, in order to evaluate the effect of yaw on entry, the air tail had been set at a slight angle which resulted in the torpedo – which had been dropped a little too early – flying in a graceful curve straight towards the launch. With a complete disregard for all his pre-flight briefing, the Squadron Leader began to pour a barrage of curses at the luckless pilot over the radio. The others watched in horror as the weapon swung towards them, dived under the boat at the last moment and shot out of the water on the other side in a cloud of spray. Other trials of a more successful nature followed, not only at the sea test ranges but also a good deal farther North.

Running trials, to establish the ability of the torpedo to maintain a true course, were held on a Scottish loch, where the torpedo was given positive buoyancy and a dye marker to indicate its position. The run was several miles in length and it frequently involved a race between the recovery team and the local Scots, who would try to arrive on the scene first and spirit the weapon to a secret hideaway in the bank. They would then sit tight and wait for the notice of a reward to be posted before producing the torpedo to the authorities.

It was while transporting one of the weapons to the loch that another incident occurred. The torpedo was collected by an armed guard from Rotol and transported on an RAF low-loader called a 'Queen Mary', the weapon being screened from prying eyes with a NAAFI tarpaulin. They made a stop for lunch at a pub near Knutsford, where they parked the huge vehicle on the forecourt. When they returned from lunch, the 'Queen Mary' was nowhere to be found and after some hectic police enquiries it was eventually discovered, safe and sound, a few miles away. One can imagine the acute disappointment when the thieves found that the supposed load of NAAFI goods turned out to be nothing more worthwhile than one of the country's most secret weapons! It would be interesting to learn the eventual fate of the armed guard when their CO heard that they had allowed their charge to be spirited away from under their noses. Perhaps they are still serving their sentence . . .

In order to straingauge the propeller shafts during entry into the water at 150 knots, the weapon was sent to Coulport in Scotland, (not far from the present nuclear submarine base at Holy Loch) where there was a facility to fire it down a ramp under the impulse of up to fourteen rockets. Straingauges were fitted to the propeller shafts, with the wiring wound round the inner and outer shaft, and trailed backwards to the shore. Although it may have appeared primitive it was perfectly adequate for the tests. The torpedo was placed on the carriage at the top of the ramp and the rockets ignited. Away it went, and at the end of the 50 ft ramp a speed of 150 knots was achieved. A good set of readings of the initial acceleration were obtained, until the wiring broke some seconds later. This was the first time that such a measurement had been attempted, and the team felt rather pleased with their effort.

Despite all the successes, the project was finally abandoned sometime in the mid 1950's. All in all, twenty four prototypes were made and a complete set of production drawings to Admiralty standards was produced, for a fully operational weapon using a sea-water battery. Terry Elmes gives the following version of the closure of the Pentane Torpedo Department:

> The end came as far as Rotol was concerned when Sir John Evetts came over to see me and announced that we were to be disbanded to provide a boost in manpower. He pointed out that the order books for aircraft equipment were full to overflowing and no more labour was available in the area.

Subsequently, the Admiralty were believed to have implored Rotol to reconsider, even offering a pre-production contract for 79 torpedoes at around £10,000 apiece but the Board were unmoved. The company severed all connection with marine matters until the Dowty take-over, when a fibre-glass boat powered by a car engine driving a water-jet unit looked like being a world beater, but eventually was to suffer the same fate as the Pentane torpedo, all of which lies outside the scope of this present history.

The Rotol Sports and Social Clubhouse in the 1950s.

The spartan appearance inside the clubhouse was typical of the times.

Queen Elizabeth viewing a Rotol propeller for the RR Dart engine at an exhibition of local industry at Gloucester in May 1955. Alec Wills of the Sales Department is on the right.

L.G. Fairhurst, Rotol's newly created Technical Director, explaining a detail to Air Marshal Tedder (Chief of the Air Staff), during a visit in 1956. Sir John Evetts, Managing Director is looking at the camera and Colonel King, Works Director, is on the left.

Rotol and British Messier exibits at the 1956 SBAC Show.

Contra-propeller for the Eagle-powered Wyvern in a test hangar installation *c.* 1949.

Ted Edwards (with folded arms) and other technicians, stands in front of a new Italian transport aircraft, the Breda-Zappata BZ308 fitted with Rotol five-blade propellers. August 1948.

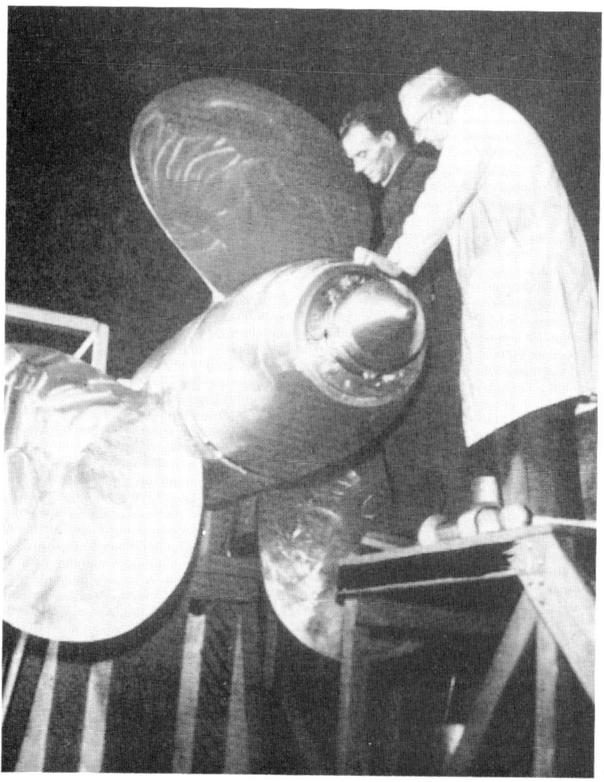

Bill Murfitt and Lloyd Jones (in white coat) check a marine propeller destined for the US Navy's USS *Mills*.

John Boast (second from right) and his draughtsmen in 1956. Eric Stroud, who served the company for almost forty years, is on the left and the author is far right.

Let battle commence! Hectic activity at the Main DO Christmas party. Team members on the left include Jim Wall, George Bubb, 'Dan' Danvers, John Russell.

In September 1954 Lord Hives, Chairman and Joint Managing Director of Rolls-Royce, opened the new Training School at Rotol.

A view of part of the workshop on the ground floor. (The drawing office was on the first floor.)

Instructors outside the recently opened Apprentice School in 1954. Seated: Green, Brown, Peebles, Faulkner, Cuthbert. Standing: Griffiths, Stanton, Howells, Clutterbuck, Dennison, Bunkell.

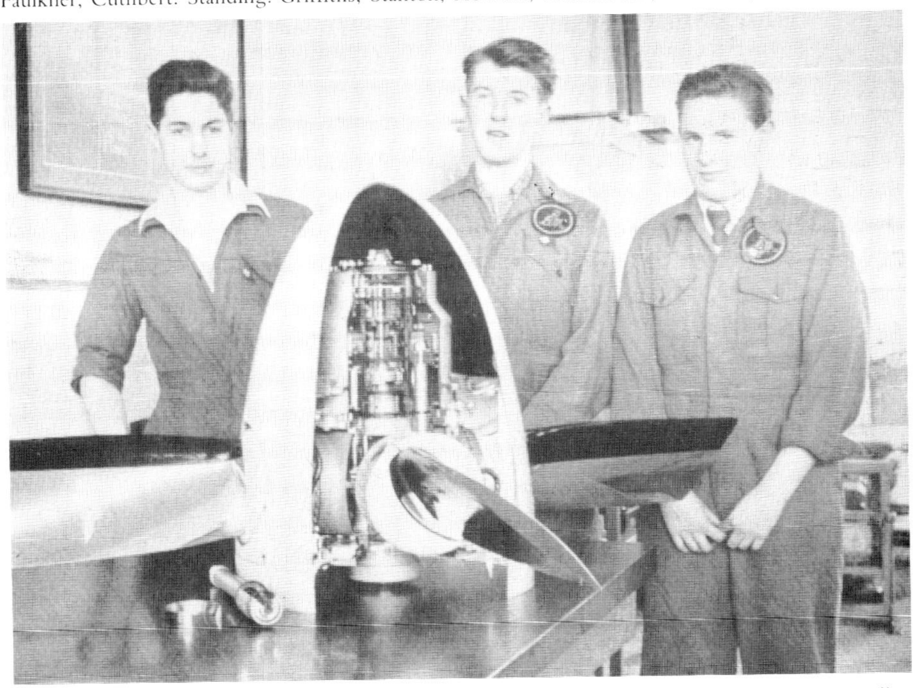

Three Rotol apprentices, J.D. Floyd and A.R. Slatter with a sectioned model of a Viscount propeller hub in 1955. Dave Griffin, right, is still with the company at the time of writing.

Rotol Flying Club members Bob Bass (left) and Squadron Leader Ken Daniell, Chief Flying Instructor, carry out a pre-flight inspection.

Jack Meaden, chairman of the Rotol Flying Club, with the Auster, here rebuilt as an Alpha.

Ray Base is the licensed aircraft engineer.

Sport and other Entertainment

Organised sports at Rotol are believed to have begun soon after the company began to operate from the Staverton site, probably assisted by the Welfare Society who were the forerunners of the present Sports and Social Society. At the outset, cricket and football matches were played on the field alongside the factory, where the present car park is situated but early in the war the company took over the nearby farm and buildings. The tenant complained that 'the farm holding had been ruined for agricultural purposes by war protection work carried out by the RAF, the Army and Rotol.' This discussion had taken place at the Directors' meeting in October 1940, when consideration into future use of the farm was given. These included conversion of two cottages for workers and turning the whole of the farmstead into a first class Welfare Sports Ground. The exact date when these changes took place has not been established but the Sports Field in its present location has been in operation for a good many years. The farm buildings, used at one period as a Riding School by the Sports and Social Society, were finally demolished in the 1970's.

The first secretary-manager of the Sports and Social Society was W. Bilton, who had joined Rotol in 1937 as Ambulance Room Superintendent. As an ex-Royal Navy man he was called up for war service and came back to Rotol in 1945. He occupied the post until his death in 1955 and was succeeded by V. Baguley, who had also seen military service. Vic insisted that when the German Army invaded Norway in 1940, he had the distinction to be the last British Officer to leave. He will be best remembered for introducing the Rotol Prize Draw in 1955 as a means of raising extra funds for the various sports and social sections run by the Society, which at that time numbered twenty one, ranging from Angling to Tennis. In its first year of operation, £4,737 was paid out in prize money and the draw itself provided an income of £1,537. Membership of the Society cost twopence per week in 1956 and the total subscriptions amounted to £1,787. The Prize Draw has continued without interruption since that date and over the years many thousands of pounds have been paid out in prize money.

The Sports and Social Society was responsible for all the various sections which were created according to popular demand. They are far too numerous to mention individually but I have singled out a few in the narrative. The Society also took charge in organising the Annual Sports Day and the Children's Parties at Christmas. The first Horticultural Show and Sports Day at Rotol took place on September 12th 1942, in 'ideal weather conditions, which, combined with the splendid exhibits and varied athletic events to produce one of the most enjoyable days spent for a long time', according to *The Airscrew*. The whole day was packed with events, as apart from the horticulture and sports, a boxing ring was set up on the field and the final of the Rotol Inter-departmental cricket competition was won by the Main DO. In the evening the Club House was thrown open to unspecified social activities, which appear to have consisted mainly of some form of concert.

The following year, F.C. Bunn headed a committee to organise the annual Horticultural Show and Sports Day. It was acclaimed a great success, with many entries in the flower and vegetables class and the day was once more favoured with grand weather. M. Howells of the Tool Room won the 'R.H. Coverley Challenge Cup' for the second successive year, with 10 First, 8 Second and 5 Third Prizes for his garden produce. No less than nine cups were presented for the athletic events, and the name of Tom Price was prominent among the winners – apart from the Men's High Jump, and Putting the Shot, Tom also won the Pillow Fight!

In the 1950's the formula was expanded and the Rotol Sports Day became a major event in the local athletic calendar and competitors came from many parts of the Midlands and the West Country to participate in the open classes. The competition was run under the auspices of the Amateur Athletic Association and usually resulted in some record times being set.

The secretary-manager had a small staff to help him run the society, among them were Maureen Stafford (née Trapp) who came to Rotol in 1955 as a young girl, initially with Tom Cruwys in the Publicity Department (where she was employed as a junior artist) but after a short while, because of a cut-back, Maureen was offered a post with Vic Baguley, taking over from Betty Griffiths (née Drake) who was leaving to start a family. Maureen stayed with Rotol until 1969, when she left for the same reason. In more recent times she has immersed herself in politics, and in 1988 had the distinction of becoming Mayor of Cheltenham. Maureen was not the first Rotol employee to occupy this high office, as Kirry Frewin, a Methods Engineer who retired from the company in 1977 after serving nearly forty years, became Cheltenham's mayor in the same year.

Two of the most popular sports, for which sections were formed at a very early date, are Rugby and Cricket. During the war, great stress was laid on sport as a means of unwinding from the hard toil of the day – or night. Rotol Rugby Football Club was founded in 1940, with Mr Stammers as President and E.F. Martin the secretary for the first season, followed by S. Bartlett until the RAF claimed his services. Despite unavoidable changes, due to members being called up for service in HM Forces, they fielded a most formidable team; by the middle of 1942 they had played a total of 41 matches, won 32, lost 5 and drew 4. The team went on to still greater heights in the 1941–42 and 1942–43 seasons when they did not lose a single match, which has to be something of a record. A note in *The Airscrew* relates that on February 17th 1945, Rotol lost to Newport by one point, a drop goal to a try. It was the first time that Rotol had lost at home in five years!

Many of the players continued working at Rotol long after their playing days were over, and one, Tom Price who has already been mentioned, played for England and toured Australia with the Barbarians and eventually had a street in Cheltenham named after him. Tom was a member of the Main Works Fire Brigade Team, which won every event in the Brigade Competition on September 16th 1944. The captain of the Rotol XV was Bill Warner, who worked in the Repair Department during the war, and was known as a man of iron. During one match he received a severe blow to the head which necessitated medical attention and twenty stitches, but insisted on returning to finish the game! Do they still make players like that? Other names from the halcyon days of Rugby at Rotol that will bring back memories were Herbie Rickards, Bill Jones, Stan Walden, Phil Pryse, all of whom – with many others – were a part of a wonderful team, long remembered.

It is probably not surprising to find that many of the names associated with the Rugby Club can also be found in the list of cricketers. In the summer of 1941 we see that young men who had been stalwarts in the mud of the previous winter were now taking wickets and scoring runs, both at home and away, despite the little spare time which was then available. On September 6th 1941, Jack Pickford, the captain, took 5 wickets for 27 runs and Tom Lacey 4 wickets for 21 runs against the Wagon Works CC and D. Medcroft scored a useful 35 with W. Perry 55 not out, giving the Rotol side an easy victory. In the Inter-Departmental Cricket Competition held that same summer, it is gratifying to the author as an old draughtsman to note that the Main DO beat the cup holders, the Maintenance Department, in a game played on a fine evening with a good crowd present. The Maintenance boys were all out for 29 runs and Main DO made 44 for 7. Jack Pickford (once more the captain) received the 'Mrs R. Stammers' Cup from the lady herself, and in the following year the Main DO won the cup outright. Over the years a great many employees have played cricket at Rotol and it would be obviously impossible to name them all but a number of people have spoken of Charlie Davey - always opening bat – and Bob Jennings as being worthy of note.

This is an appropriate moment to mention the name of another Rotol man (his name has already been quoted in these pages in other connections but we must include it in this section in the name of common justice) none other than L.G. Fairhurst. He seems to have played cricket and football for a great deal longer than most, although not without incident. He was something of a demon bowler in his youth, although he seems to be best remembered for his batting prowess in later years. In *The Airscrew* of August 1945 there is an account of a match between the Executives and the Rotol Club XI which sounded most enjoyable, even if the dream of the 'old hands' giving the premier side a sound thrashing was not realised. The Club declared at 102 for 6 – at one time they had been four wickets down for only 39 – but apart from a useful 27 from George Bubb, the Executives did not manage to shine and were all out for a lowly 58. Special mention was made of Reg Barrell's wicket-keeping (he conceded only two byes) and Les Law took three grand catches. Messrs Jelley and Mangeot were best with the ball and Bubb and Fairhurst took the batting honours. The Rotol Executive Staff played an annual match against the Bristol Aeroplane Company with LGF as captain and latterly Charles Bracher, until the matches eventually petered out sometime in the 1970's.

Ian Golesworthy recalled how he persuaded LGF to play for the Technical Block side in the finals of an Inter-Departmental soccer match in the early 1950's when LGF at 49 must have been the oldest player in either team. Unfortunately, he was ceremoniously carried off in the last fifteen minutes, due to injury and subsequently had to undergo twice-weekly physiotherapy for the next two years. John Chilman, who is about five years younger than LGF, also received a tendon injury in the same match, and for both men it was the end of their days on the football field. At the next meeting of the Rotol Board of Directors LGF was sternly criticised by Hives for foolishly exposing himself to the rigours of the sports field at his age. In correspondence with the author, LGF remarked in a poignant postscript, 'I was warned many years ago after sustaining a broken nose, to give it up as I would only suffer in later life – but "later life" seemed so far away.'

Jock Ancell reminds us of another tale concerning a cricket match that took place over fifty years ago, on the original sports field where the car park stands at the time of writing. It was 1938 and the first time the final of the 'Stammers Cup' was played, when Jock's team were all out for 24 runs, of which he made 15 and the Senior Staff side (including LGF) were 'skittled' out for a miserable 12! He maintained that the pitch was pretty poor, which accounted for the low scores.

While the Rotol male elements were busily engaged with their rugby, football and cricket, the ladies seem to have opted for two sports which were just as strenuous but more usually associated with the fair sex, those of tennis and hockey. The former, the Rotol Lawn Tennis Club, operated from two grass courts at Barnwood Road and two hard courts at 127, Old Bath Road in Cheltenham. The committee in 1942 comprised one lone male, Mr Conway-Gwilliam, the Chairman and eight ladies, among them Miss Florence Jackson, the Controller's secretary.

The Ladies Hockey Club was formed early in the war (a brief reference to matches played in 1939 seems to confirm this) but it is not until the winter of 1942/43 that we begin to see photographs in the magazine and are able to follow their fortunes. The indifferent weather conditions and other unavoidable circumstances were the reason for many matches being cancelled but the ladies could take satisfaction from the fact they won 9 out of the 12 games they played. On one occasion the Rotol girls beat a WAAF team by 23 goals to 1 – and they were playing one short! The centre-forward up until this time was Miss Gatfield who left shortly afterwards to join the ATS and her absence was keenly felt for a while.

One aspect of sporting life at Rotol cannot be overlooked as it formed a major part of most recollections among the employees, and that was boxing. The Rotol Amateur Boxing Club held tournaments in the works canteen during the lunch break, throughout the war and occasionally at RAF stations. (The magazine was not permitted to name the location but presumably it was either RAF Innsworth or RAF Honeybourne but there were many others in the area.) These matches were always well attended and reported on at great length in *The Airscrew* together with many photographs of the contestants. Boxing continued to figure at Rotol well into the 1950's but has never been revived in recent times. Perhaps the sport has lost some if its attraction in the present climate of public opinion.

Turning aside from the more conventional sports at Rotol, we will survey the other activities that the workforce enjoyed, either as individuals or, more commonly, as a section or group. Both Rotol and British Messier had a small nucleus of people in almost every department who would organise an event in which all could participate, whether it was an informal skittle evening, a dinner, or a party.

The Rotol Technical Block Christmas Party first took place in 1949 and continued without a break for many years but died out after the Dowty takeover. It was confined to those in the Technical Block which included the Main Drawing Office, Marine DO, Project Design, Project Stress, Performance and Stress Offices, and the Prototype Department. (My apologies for any I have inadvertently omitted.) Secondly, although there were a great many ladies in the block, they were not invited – it was strictly a 'Stag' occasion. The reason for this is not difficult to find as the first part of the evening was devoted to a fairly rumbustious and hectic period of games, and although it may be fashionable nowadays for the fair sex to take part in everything which was once considered a male preserve, in those days it was not. The preparations usually began soon after the summer holidays were over, when a small committee was formed to plan the event and regular meetings were held. In the beginning this comprised, John Hellier, Alan Walker, Roger Armstrong, Eric Langton, Wilf Gordon, and Ian Golesworthy who had joined Rotol in 1950. Others were called upon from time to time and each of the separate offices were encouraged to provide entertainment to conclude the evening. A programme was produced, initially merely a typewritten affair but later by offset litho, courtesy of the Technical Illustrator's Department. This detailed a list of the teams, times of events and rules for the games – although this was never allowed to interfere with ones enjoyment. It also contained a number of sketches from the gifted pen of Ray Sherbourne, a draughtsman with a flair for cartoons, who sadly died at an early age, a few days before Christmas 1964.

Before the evening revelries began it was customary to have a meal in the canteen, and those who were attending the function left their offices half-an-hour before finishing time. The Rotol Clubhouse at that time was situated where the Bowls Pavilion now stands (a serious fire destroyed a great deal of it fairly soon after the new Clubhouse was built in the early sixties) and was reserved exclusively for the use of the Technical Block Party for the occasion. The six or eight teams were comprised of members from the same office or section, usually about a dozen strong, depending on the numbers attending. Various games ensued, Indoor Cricket in the skittle alley, Blow Football on the table tennis table, where, according to Ian Golesworthy, 'the great bulk of the Chief Designer, 'Dan' Danvers dominated the proceedings like a hippopotamus in rut'. Dan was an immensely powerful man, with something of a weight problem and is reputed to have been restricted to eating raw meat in order to control his diet – with what success is not known. During one of the Christmas parties he proved his great strength by lifting an opposing team member clean off his feet and holding him over his head.

Another enjoyable and vigorous game was played by two teams facing each other on chairs, with rolled up newspapers and a tennis ball. It was supposed to be a version of hockey but over the years it became known, by virtue of its main ingredient, as 'ead 'ittin'. The ball would often disappear for minutes on end but the game went on! Points were awarded for all of the games and the winning section received a rousing cheer – although in the great British tradition, 'It mattered not who won or lost but how we played the game'.

After these strenuous pursuits were over everyone assembled in the lounge and settled down for the remainder of the evening, an entertainment of sketches or songs, loosely based on events of the past year and featuring some of the main characters in the factory and offices, together with slightly libellous accounts of the activities of the heads of departments and higher management. The 'entertainment' had been written and rehearsed for many months and each item was greeted with enthusiastic applause by the assembled company. Even the 'chiefs' entered fully into the spirit of the evening, either taking part in the sketches or 'doing a short turn' on their own. The raucous singing of the traditional air 'Auld Lang Syne' rounded off the evening and the revellers departed, ready to do it all again next year.

The Technical Block Christmas Party was the highspot of their social calendar but this is perhaps an appropriate moment to briefly mention some of their other activities. When the sun shone in the summer months – as it sometimes did in those far-off days – Ian Golesworthy would organise the very few car owners in the offices to take a mixed group to the Cheltenham Lido for a quick swim in the lunch hour. (The Lido was also very popular with other groups during the lunch break, among them the men and women of the British Messier technical offices.) Ah! the mad dash along roads that had hardly any motor cars and practically no traffic lights, just to get about thirty minutes of refreshing cool water on our bodies. How we survived the rest of the afternoon I cannot remember . . . but we were all so very young then.

The members of the Structures Office were always ready to hire a coach and 'get out on the town'. With sixteen of the fair sex and only nine males, it was no wonder that they insisted on any team which challenged them to a game of skittles being composed of a similar mix. The Red Lion at Wainlodes was one of their favourite haunts, as it was for so many of the other departments at Rotol. Everyone who recalled the Red Lion seems to have had a story to tell of one of the more inebriated members of the party being involved with a swim across the Severn, at sometime during the evening – the author recalls that Ian Golesworthy often seemed to feel an urgent need to get into the river at the end of the evening.

When Charlie Harmer joined the Flight Test Department in July 1944, the total work-force comprised some ten riggers, ten fitters, eight female trainees and three pilots. Sadly, Charlie was one of those who passed away during the time that this book was being researched, but he recalled many amusing anecdotes from the period and laid particular stress on the excellent social occasions and considerable amateur talent available, enabling them to hold regular party nights in the old club house. One of the fitters, John Spinner, was a clever impressionist who could impersonate many of the well-known personalities of the period. Walter Gainey, an electrician, could do a wonderful stand-up comedy routine, sometimes lasting up to half an hour. Walter was a man of many talents, being a highly qualified member of the British Red Cross and also an expert at shooting and fishing, all of which gave him plenty of material for his comedy act.

On one occasion, enough talent was found to form a six-piece dance band, and St John's church hall was hired in Churchdown, where their music was much appreciated by the members of the RAF and WAAF who turned up in strength from the nearby Innsworth camp. In addition to all this wealth of musicianship,

Charlie paid special homage to Ted Edwards, who led an excellent vocal section and people from all over the factory came along to enjoy the parties which Flight Test gave in the old Rotol club house. No doubt these all served to keep up morale during the war years, and were a welcome break from the long working hours and the bleakness of the times.

The pages of *The Airscrew* and the *Rotol and BM Review* abound with details of the various dinners and parties organised by the machine shops and offices during the Christmas period. The examples which follow, all culled from the magazines, may be considered typical for almost any year.

One of the most prestigious of the dinners given during the festive season was the Supervisors' Dinner, traditionally held in the Canteen, where frequently over 300 sat down, with distinguished guests and their wives at the top table. The Planning Department usually held their Christmas dinner at the Rotol hostel, where Olde Time and Modern Dancing formed the entertainment for the evening.

In the 1950's, the Gearbox Shop celebrations in the Clubhouse at Christmas was a rather more serious affair than already related, with employees, their wives and friends all taking part. Games of skittles and cribbage were organised and the evening was rounded off with spot-prize dancing. Phil Pryse won one prize and Eric Searles another, and there were ugly rumours of the whole thing being 'fixed'! It was so successful that the event had to be staged on two consecutive evenings in December.

Among the workers at the Repair Department in Gloucester during the war was Bill Warner, best remembered for his prowess on the rugby field. He and his friend Jim Phelps were always at the hub of any parties being organised at ARD and both of them are mentioned in this connection in several editions of *The Airscrew*. One copy carried a report on 'the Annual Turkey Dinner, when over one hundred people sat down in the canteen to a most excellent repast provided by Mrs Kirkwood and her staff, followed by musical and humorous items from members of the ARD.' This particular party occurred on January 7th 1944 but both Bill and Jim had already been responsible for a party on December 23rd 1943, held in one of the local pubs – where there was the usual combination of wine, women and song! Although there had already been much jollification during the lunch-hour of the same day, this was an evening get-together, when Jim Phelps, the compere, introduced the talents of several members of the repair organisation. These included Mona Raye who sang and led the audience in the choruses and George Trimby who played two 'hot' numbers on the piano. One of the biggest laughs came in a comedy duo act, when Jim Phelps asked, 'What do you think is the finest job Rotol has ever done?' Arthur Redmond replied, 'When the designers put their heads together and produced a wooden blade!' (Eat your heart out, Tommy Handley . . .)

The first performance of the ARD Concert Party had been held in the Glevum Works Canteen on August 18th 1943. They were scheduled to take place in the lunch break every Wednesday during the winter months and seemed to have a wealth of talent from within the ranks of ARD. The versatile Arthur Redmond was an illusionist of some note with his burning Postal Order trick, and Gordon Edgar evoked roars of laughter with his Yorkshire humour and Anne Waugh sang 'Birdsong at Eventide'. F.L. Valentine (Val) acted as the Stage Manager, and in this capacity he approached a local undertaker to enquire if he might borrow a silk top hat. This was accomplished but when he went on to enquire if they could oblige with the loan of a one-string fiddle he was told that one-string fiddles were not in the undertaking line but if he should require a harp on a future occasion they would be happy to accommodate him.

Rotol always seemed to have talent in abundance for almost any activity which might be mentioned. One which was predominant during the war years was that of entertainment for the workers, which usually took place in the Canteen during the lunch-break, to the accompaniement of a clatter of knives and forks. The man who was mainly responsible for almost all the many stage shows, several of which were broadcast, was Ron Brown. He was a member of Bert Jelley's staff in the Works Engineers, where his work covered 'passive defence measures' – a term covering the wartime shelters, first-aid posts, plant and fire protection, both at Main Works and dispersal areas. The pages of *The Airscrew* carry reviews of the many concerts organised by Ron Brown and he and Tom Cruwys performed a double act which featured on several occasions.

The canteen was not only used for concerts during the day but also, on rare occasions, in the evening too. A 'Star Variety Programme' took place on December 7th 1948 at 7.30 pm in the canteen, when Frankie Howerd ('Ladies and GentleMEN!') was top of the bill and just over a month later Dickie Murdoch and Kenneth Horne appeared. Both of these concerts were organised by Ron Brown.

The Rotol Amateur Dramatic Society was also well to the fore but they usually put on their productions in places such as church halls in the locality. *The Airscrew* of June 1941 in reviewing the play 'Busman's Honeymoon', tells us that they drew a good-sized audience to the Holy Apostles' Church Hall in Charlton Kings. It was their first performance and among the cast were several people who have already been mentioned in these pages; Ken Thatcher took the part of Lord Peter Wimsey and was also responsible for the production. Also in the cast were Jack Pickford, Norman Robertson and Joan Hart (soon to become Mr and Mrs) and 'Pop' Milner, whose portrayal of the Reverend Goodacre was – 'lifelike to the point of inaudibility', according to the review by John G. Walter. They must have been encouraged by their first venture as the RADS went on to put on several more productions.

Rotol at Peace

After the undeniable success that Rotol had enjoyed up to now, it is sad to relate that the first post-war propeller design was something of a failure. Although, in the last year or two of the war, the two-blade constant speed propeller for the Avro Anson had been a great success with over 3,000 being manufactured, an attempt to produce another two-blader did not meet with the same acclaim.

The 'Junior' Rotol 6 ft diameter, two-blade controllable pitch and feathering propeller, was announced in 1946. It was intended for light aircraft powered with engines of 80 to 150 hp with the blade angle controlled by the pilot, who could select one of three positions, 'take-off and climb', 'cruising' and 'all-out level flight', or any intermediate position of his choosing, to suit blade pitch/rpm condition. An electric actuator, connected with a flexible drive to a screw jack, changed pitch through a mechanical linkage. An alternative version for single engine aircraft was also proposed, in which a hand crank could be used by the pilot. The cost was anticipated to be in the region of £100 or £120 depending on the method of actuation. It may have appeared an attractive proposition, but there was insufficient interest shown and the propeller never went into production. Thirty years later, Dowty Rotol, under their new technical director, John Keenan, resurrected the principle of cheaper propellers for the light aircraft market, which went on to eventually enjoy some success.

By 1946, both Rolls-Royce and the Bristol Aeroplane Company had brought piston engines to a fine state of the art, and although this was probably not apparent at the time, versions of the Merlin, Griffon, Hercules and Centaurus were to carry on for several years. But a great deal of uncertainty was present within the aircraft industry, and in particular a large question mark hung over the future for propeller-driven aircraft. Many 'experts' were predicting that the turbojet engine would supersede all other forms of propulsion, and propellers would only be used for light aeroplanes.

There is a possibility that with the war over and times undoubtedly changing, the Rotol Directors may have tended towards this viewpoint, and no longer had quite the same enthusiasm for continuing with the propeller business. The parent companies original aim, of ensuring that Britain would be in the forefront of propeller development, had been successfully fulfilled and other matters now claimed their attention. Gas turbine engines to power the aeroplanes of the future were the main concern of both parent companies, and a great deal of their efforts went towards this development. At the forefront of the Directors' minds must have been the question of deciding if there was still a large enough market for aircraft propellers and how big would be Rotol's share?

If there was any uncertainty on the part of the Directors it was not held by L.G. Fairhurst, the Chief Engineer. In July 1946, convinced that a secure future still lay ahead for certain types of propeller driven aircraft, fully recognising that many military aircraft would inevitably take advantage of the higher speeds offered by the jet engine, he produced a paper outlining his view of the future, including propellers driven both by conventional piston engines and gas turbine power. Despite whatever doubts the Directors may have had, both the propeller and auxiliary gearbox business went ahead, albeit on a somewhat reduced scale, and many looked back with a feeling of regret to the days when Rotol were delivering huge quantities of both these products to engine manufacturers and maintenance depots all over Britain.

At the end of June 1946, L.B. Greensted, who had been awarded the MBE in January of that year for his test flying duties, made his last flight with Rotol and resigned from his position as Chief Test Pilot, joining the newly-formed Skyways Limited as Chief Pilot. Two years later Skyways were to become very heavily committed to the Berlin Airlift and Greensted was appointed Skyways Berlin Airlift Manager in 1948. This was not just a desk job, as he recorded a personal record of 325 flights into Berlin (Gatow) delivering over 3,000 tons to the beleagured city, for which he was awarded the King's Commendation for Valuable Services in the Air. Space does not permit more than a brief glance at his subsequent career but his involvement with the introduction of the Vickers Viscount and the Bristol Britannia into service by the independent operators was considerable. As Chief Pilot and Operations Director of Hunting-Clan Air Transport, he was responsible for all operational staffing and procedures for the Company, including cabin staff and maintenance standards. It was during this period that the Viscount was introduced into scheduled services throughout Africa and the Britannia on HM trooping services to Singapore and Hong Kong. The post of Chief Test Pilot for the Rotol Flight Test Department went to Squadron Leader Kirkland, who had been with them for some months.

With all the changes of personnel taking place, one unexpected vacancy occurred when Eric Cameron died suddenly in 1946. He had held the position of Company Secretary from the very first and his loss must have been keenly felt. Within the space of one year for a variety of reasons, the company had lost its three top men, Coverley, Stammers and Cameron. An ex Rolls-Royce employee, D.A. Tomlinson, was appointed company secretary.

One man who has not so far received any prominence, although he was with Rotol for as long as anyone could remember, was Charlie Homer. A well-respected and rotund individual, with several supervisory appointments, Charlie was a familiar figure throughout the factory. Phil Smith, later of the Works Engineer's Office, was an apprentice in the early years of the war and recalled that Charlie had an office built on stilts, somewhat akin to a guard tower, so that he could have a good view over No 1 Machine Shop. This was not too popular with the workers and they took a surreptious delight in bombarding it with an occasional missile from a safe distance, causing Charlie to stand up and glower over his spectacles across the shop. Phil does not admit to this pastime but remembers that as a young apprentice, 'if ever Charlie smiled, it was never at us'.

Doug Grazier was another well-remembered personality in the company. He came to Rotol in 1946 as a fitter in the Maintenance Section and contributed the following example of Charlie Homer's particular brand of humour, when Charlie was the Superintendent in No 1 Machine Shop:

> I was asked to repair a Warner and Swasey turret lathe in No 1 Shop and when I turned up, the machine was completely stripped down and all the gears, bearings and spindle removed and dumped in the suds tray. There appeared to be no valid reason why this had happened, so I asked for an explanation from the Senior Foreman Bill Edden, who walked off to have a discussion with Doug Culliver the Line Foreman, and they were soon joined by Charlie Homer. After a while, Charlie called me over and asked for an estimate of how long it would take to put the lathe back into working order. I gave him an answer and he nodded thoughtfully and said, 'Come to my office in ten minutes and bring a large screwdriver with you'.
>
> When I turned up, Charlie had Bill, Doug Culliver and the operator – who will remain nameless – in front of him and was making small talk in a friendly manner. Charlie casually enquired the time and the operator proudly showed him a very nice watch he was wearing, taking it off his wrist and passing it to him. 'Lend me your screwdriver', said Charlie, turning to me, which he then applied to the back of this expensive watch. 'What the hell are you doing?' shouted the operator, leaping to his feet in horror. Charlie's eyes gleamed, with a look which boded ill for the luckless individual standing before him. 'Nothing different from what you have done to my machine', he replied. At this point I was asked to leave the office and therefore missed the next few minutes, when Charlie delivered a typical 'Homer Special' or as we called it – a wire brushing – of monumental proportions.

Doug Grazier continued with the company in various positions of responsibility until his eventual retirement in 1986.

The largest landplane ever built in Britain, the giant 230 ft span Bristol Brabazon airliner, which first flew on 4th September 1949, also had the largest diameter propellers to be made by Rotol. No less than four pairs of Bristol Centaurus engines, served as the power units for this immense aircraft, each pair geared together by an arrangement of internal bevel gears, driving the co-axial shafts on which were mounted the 16 ft six-blade propellers. The aircraft was without doubt a bold venture, and the airscrew design that had been commenced in late 1943 by the Rotol technicians was a triumph of engineering. In addition to controlling propeller pitch for constant speeding, feathering and braking functions, the system introduced synchronous control of the engines. By this means, with one of the engines chosen as the master the other seven could be synchronised to the same rpm thus largely overcoming excessive noise.

Unfortunately, the Brabazon was never to be built as a production aircraft and although it was always marvelled at by those lucky enough to see it in flight, it suffered the fate of many other great aeroplanes, being reduced to scrap in the early fifties. However, Rotol continued to supply propellers for a number of other aircraft produced by BAC in the post-war period. These included the Buckmaster which served in the RAF until 1955 and the Brigand, produced in 1950 and remaining in service until 1958. One of the last propeller-driven Bristol aircraft to have Rotol propellers was the Wayfarer which made its first flight in December 1945, and although Rotol produced a number of four-bladed propellers for the early versions the bulk were supplied by De Havilland's, who also produced the propellers for the much larger Proteus-engined Britannia.

Almost the last 'modern' aircraft to have wooden propeller blades was the original Vickers Viking, the first British post-war airliner, although they were only fitted to the prototype aircraft. The production Viking, and the more numerous versions developed for the RAF as the Valetta and Varsity, were fitted with Rotol solid Duralumin blades. Jock Ancell recalled the time when the whole of the BEA Viking fleet was grounded due to a problem with the propeller, caused it was thought, by the blades becoming loose. He and Ted Burchill were given the task of trekking all over Europe in the depths of winter to fit brand-new propellers to the stranded aircraft. One of the aircraft was in Vienna, in the Russian occupied zone and it was the coldest occasion he can ever remember.

This tale echoed one which occurred in 1941, at the time when a number of Hurricane fighters, fitted with Rotol propellers, were sent to Russia. The extreme cold had loosened the wood blades in the metal adaptor and – so the story goes – the Russians got round the problem by simply drilling a hole through the two parts and fitting a bolt with the end rivetted over. Unlikely though this sounds the evidence can still be found in the DO records. In order to regularise the *fait accompli*, RA.11380 a ⅜ BSF 'safety bolt' was drawn by Ivor Brooking in January 1942 and RA.11381 showed its application, right through the wood blade root and adaptor. Modification VP363 was authorised by the MAP on January 21st 1942 and subsequently extended to cover not only the Hurricane but also the Fulmar, Barracuda and Sea Otter.

As a result of the 1945 changeover to metal blades, enormous stocks of wood blades became redundant almost overnight and stories have been told of them being sold for two shillings (10p) each for firewood, farmers using them as fencing and of fields in out of the way places being filled with heaps of these now unwanted items. Don Bishop remembered that Hordern-Richmond were selling replacement wood blades to the Ministry at fixed contract price and buying them back as war surplus for 9d. (4p) each!

The period from the end of the war to the start of the post-war recovery in Britain took some time. This was not the land flowing with milk and honey to which many had looked forward during the dreary years from 1939 to 1945. Raw materials were still scarce and food and clothes rationing continued and even got worse in some instances. Most people are surprised to be reminded, if they were present at the time, that among the cuts imposed was the halving of the cooking fats ration and the reduction of bacon from four ounces to three. After six years of war, the Labour Government was battling hard to bring the National Economy within bounds, there were all manner of shortages at home and in almost every area of British life the mood of many of the people was of a great weariness.

These were hard times for everyone, coupled with the beginning of the Cold War with Russia and cold winters too. The winter of 1947/48 was the worst of the century and the snow and ice made life a miserable existence for weeks on end, and the subsequent thaw brought even further hardships. During February and March there was a coal crisis and this, coupled with the extreme cold, made it impossible to heat the Rotol workshops and offices adequately. Everyone on the shopfloor worked in heavy overcoats, scarves, gloves and balaclava helmets. Arthur Gunn and his mates rigged up a false ceiling from cardboard above the marking out table in the Experimental Shop, to try to retain a little heat. Norman Grinnell was on the night shift during this grim time and considered it to be 'no joke', surely the understatement of the year!

With a gradual return to the better weather as the months went by, life in Britain slowly improved. At Rotol, following the cut-backs at the end of the war, new starters were beginning to appear on to the pay-roll. Among them was Muriel Cousins, whose first husband, O.C. (Charles) Jackson, had worked for LGF as a personal assistant but was tragically killed in a road accident. She had never been out to work before and life in a place like Rotol may not have appealed to her much but it was close to home and there were not many alternative jobs available. She started work in the cashier's office in 1947 and her first task was to pay out in the machine shop. Each of the girls in the Pay Office had a tray of pay packets, one for each of the men and women in various departments and areas in the factory. Instead of someone taking her to the appointed spot, Muriel was given a map of the Rotol site and headed in the general direction of No 1 Machine Shop, where her first encounter with the realities of the job swiftly became apparent.

When she entered the shop she thought she had arrived in Hell. On top of the noise from the machines, came the clamour of the men banging on their cabinets and lockers and wolf-whistling as she made her way down the gangway. Finally completing the pay-out she made a hasty dash for the nearest door and promptly found herself lost. An hour later, one of the patrol officers guided her back to the Pay Office where they greeted her return with some amusement. It was to be quite a while before she was allowed to forget that incident! Mrs Cousins will perhaps be best remembered as personal secretary to Sam Peebles in the Apprentice Section in the 1950's, where one of her many duties included listening to the youngsters who poured out their troubles and asked her advice.

Another new starter in June 1947 was Jack Meaden, who joined the Main Drawing Office after five years service in the Fleet Air Arm. He came on the scene at the moment that the Rotol Flying Club was about to disband, due to the fact that the aircraft had been out of action after being blown over by strong winds and the club was operating at a loss. The Club had been formed in February 1945, with Bernard Thompson BEM as secretary and treasurer, and by the end of the year the club had some 50 members. Civil flying was suspended during the war and did not recommence until June 1st 1946 and the club bought its first aeroplane soon after. This was a 40 hp two-seat Taylorcraft Model A, registration G-AFJ0, purchased through a loan which was repaid over twelve years. The first few months were difficult times and the club was in the process of being wound-up when Jack arrived. Because he was a newcomer, it took time for him to convince the committee that the club could be operated successfully, but eventually this was done. Some reorganisation took place, the aircraft was returned to service and flying recommenced in January 1949, and since that time the club has continued to provide flying training and recreational flying for its members.

In 1952 the Taylorcraft was replaced with a 100 hp Auster J/1 Autocrat three seater, G-AHCL, which was subsequently modified in 1956 to a 130 hp four seat J/IN Alpha. The Auster was operated by the club for 21 years, when it was replaced by a Rallye for the next 10 years. For the last 8 years the club aircraft has been a Cherokee, which the club still operates at the time of writing. For the first three years the club did not pay for fuel, hangarage or servicing, all of which came by courtesy of the Flight Test Department. In addition the company's own insurance company carried the club aircraft insurance, without extra charge. Flying cost 10/- (50p) an hour in 1947, which was only about one sixth of that charged by other clubs at the time and hundreds of people flew in the Auster as associate members. Even in 1959 the author recalls that a half hour flight in the lunch break still only cost about 10/- (50p) but following the Dowty take-over and with the passing years, the cost of flying inevitably increased. At the time of writing it is still less than two thirds of the cost elsewhere and the members, although fewer in number, are still just as keen.

The success of the Flying Club depends on the individual members but no one will deny that Jack Meaden, the Chairman since 1954, and secretary before that, has been the driving force for more than forty years. Jack was appointed Design DO Representative for the Flight Test Department at Moreton Valence in 1950, moving back to Main Works when FTD was wound up in 1954. He retired from the company in 1984 and has continued his association with the Flying Club ever since. Regrettably, in 1990 because of increasing costs, fewer people can afford to fly and the future of the club is less promising than at any time in its 45 years of operation.

The work of the Flight Test Department continued in the post-war period, but with a much reduced staff, following the 1946 redundancies. In June 1947 the department gave up the Staverton site and moved to Moreton Valence, where the longer runway meant that heavier aircraft were able to operate with a greater safety margin. They shared the airfield with Gloster Aircraft, who were testing a variety of different marks of Meteor and in 1951 the prototype Gloster Javelin flew from Moreton Valence. All the aircraft had to be fitted with R/T (radio at Staverton did not become mandatory until some years later) to establish communication between aircraft and Moreton Valence tower and the Flight Test Department continued with its test flying. Now that Greensted had departed, Kirkland and later R.S. Easby shared the pilot's duties, flying the latest marks of Spitfire and the heavier fighters such as the Hawker Sea Fury, which were just starting to come on the scene.

Dick Turley-George DFC joined the Flight Test Department in 1949 after a varied and distinguished service in the RAF He had been shot down in the Battle of Britain, and been catapulted in Hurricanes from the bows of merchant ships as part of the escort for two Atlantic convoys and one to Russia. At the end of the war he spent a year in hospital after being wounded while returning from a photographic mission. A spell as Chief Instructor of the Meteor Conversion Unit at the Empire Test Pilot's School was followed by service with No 18 Squadron in Burma and No 84 Squadron in Singapore. When he left Rotol after the closure of the Flight Test Department he went to Short's of Belfast where he was engaged on test flying Canberras.

Westland Aircraft of Yeovil have produced a variety of what might be considered unusual aircraft over the years and the Wyvern was somewhat in this mould. Although orthodox in outward appearance, it was to cause both Westland and Rotol a great deal of anguish over the next eight years. Westland first turned their attention to the requirements for a projected naval strike fighter in 1944, to be powered with a new design of piston engine, the enormous 2,690 hp 24-cylinder Rolls-Royce Eagle. The Wyvern dwarfed most other fighters – it was three times the weight of a Spitfire – and its eight-blade contra-propeller was the largest produced in Britain at that time. The RR Eagle, although fitted to the first six prototypes, was cancelled in December 1946 and a change to turboprop power followed. Two engines were under consideration, the Rolls-Royce Clyde and Armstrong-Siddeley Python and prototype aircraft using these engines were flown, each with Rotol propellers.

Within a short space of time the Clyde was cancelled leaving the 4,200 bhp Python as the standard power plant and in this form the Wyvern went into production, the last propeller-driven fighter operated by the Royal Navy, but difficulties with engine surge at high speed together with propeller response problems were largely the reason for a long delay in getting the Wyvern into service. Rotol developed a 13 ft diameter eight-blade contra-rotating propeller and supplied over 250 front and rear units for all the 127 Wyvern prototype and production aircraft that were built. During the Suez crisis in November 1956, Wyvern's of No 830 squadron operated from Royal Navy carriers in the area, carrying out air strikes on Egyptian airfields and other military installations in the Canal Zone and one year later they were being phased out of service. Test flying of various Wyverns had occupied many months at Moreton Valence and Wyvern VW882 was the last aircraft in use by Rotol when the Flight Test Department closed down in 1954.

The Fairey Gannet was another naval aircraft with a propeller system designed by Rotol that was to cause a few headaches, although never on the scale of the Wyvern. It had a co-axial propeller, comprising two concentrically mounted units, each carrying four blades and rotating in opposite directions. The power unit consisted of the Armstrong-Siddeley Double Mamba engine, the port engine driving the left-hand tractor front half, and the starboard driving the right-hand tractor rear half. This gave the Gannet the ability to shut down one of the engines when on long patrols and feather the appropriate propeller, thus conserving fuel and substantially increasing the range. These co-axial propellers were the last of the breed and no further development has been undertaken since the fifties. Armstrong-Siddeley also produced the Mamba engine in its single form, powering the Seamew, for which Rotol supplied the propellers.

Visits to the United States had occasionally been made by Rotol personnel over the years, mainly in either an investigative or consultative capacity, and Rotol had also received many American visitors. Over a dozen representatives of an American Air Mission toured major British factories, including Rotol, in October 1942. Coverley and Newport made a tour of propeller and aircraft manufacturers in the USA at the end of 1944, mainly to discuss the question of hollow-steel blades produced by Curtiss Wright. (LGF had been scheduled to accompany Coverley but a sudden illness that was to put him out of action for several weeks meant that Newport went in his stead.) The first Rotol employee to be sent to the New World is thought to have been T.F. Barrington, in February 1942, who was one of Coverley's Technical Assistants, but why he went and what his duties were has not been established.

Bill Templeton, assistant to George Harvey the Installation Engineer, went to the United States towards the end of the war to supervise the fitting of a five-blade airscrew to a lightweight version of the very successful American fighter, the P-51 Mustang. The propeller installation was successfully accomplished and had the benefit of considerably reducing the all-up weight of the two XP-51G lightweight prototypes but the P-51H production aircraft which followed were fitted with four-blade American propellers. (There had already been a small order of 25 airscrews for the Mustang, but these were the four-blade variety and were presumably for trial purposes).

When this story was retold to the author, it was suggested that Bill so liked the look of America on descending from the converted Liberator bomber in which he crossed the Atlantic that he immediately began enquiring about taking out citizenship. During his stay, Bill had been accepted into the Hollywood set, where he had the good fortune to be smiled on by Carmen Miranda. She proceeded to keep him in a style to which he was quite unaccustomed, but which he very quickly settled into. Unfortunately, Ivor Brooking, who is credited with this tale, does not reveal Bill's subsequent movements, whether at Rotol or at 'Tinsel Town', although it is thought he eventually took employment with BOAC in Canada.

Rotol made its first post-war sally into the North American continent in 1947. TCA (Trans Canadian Airlines) were operating Douglas DC4 aircraft, which had been modified by the Montreal based firm of Canadair, replacing the four Pratt and Whitney radial engines with the Rolls-Royce Merlin. The Hamilton Standard propellers originally fitted, had also been suitably modified for the new installation. TCA requested Rolls-Royce to step-up the engine rating by 100 bhp to permit an increase in take-off weight from 80,000 lbs to 82,000 lbs and Hives instructed L.G. Fairhurst to visit Canadair and discuss the problem with the aerodynamicists and design staff. After a fortnight, a suitable proposal was submitted to TCA but unfortunately the cost of clearing the new propellers to meet the official airworthiness requirements was considered to be too high and the project was cancelled.

In the mid 1950's, Rotol received an enquiry from the Spanish Air Force to fit propellers to Merlin-engined versions of the Messerschmitt Me109. Spain had a long association with this aircraft stemming from the days of the Civil War, and had even ordered a small quantity during the Second World War. Germany had supplied unassembled airframes and a licence to build them in Spain was agreed in 1943, but due to the Allied bombing campaign, no engines were received from Germany. The airframes were modified to take a development of the French Hispano Suiza engine, the Spanish-built HS-12 which was used in production of the HA-1109 by Hispano-Aviacion at Seville from 1945 to 1951.

Because the engine was not entirely satisfactory it was decided to fit the Rolls Royce Merlin in its place. By this time, Rotol no longer produced suitable airscrews for this installation but LGF thought an ex-Ministry store might be found that still held some stock. He approached Jock Ancell, who was considered to have a wide knowledge in this connection. He wasted no time on his quest and after locating a supply of ex-Seafire propellers, even negotiated a price for the complete airscrew of £15. There remained one final hurdle. The wooden blades were unacceptable to the Spanish authorities and had to be exchanged for metal. Once again, Jock came to the rescue, with the whereabouts of some ex-Firefly units, and after they had been 'cropped' to reduce the diameter, they were ideal. Rotol GRF/5 Constant Speed Units were obtained from a similar source and suitably modified by a small change to the Pressure Relief Valve, and resulted in a total of over 200 propellers being supplied before Spain ceased production of the aircraft in 1958. Some of these ex-Spanish Air Force machines are still flying today, suitably painted to represent German aircraft of the Second World War, and are regularly used by film companies. Readers will no doubt remember the Cinemascope version of *The Battle of Britain* made in 1967, when they were much in evidence.

The possibility of aircraft propellers lasting for many decades must have seemed highly unlikely to most people in wartime. It was only in the late 1960's that Dowty Rotol began to realise, with the enormous interest in Second World War aircraft, a basic requirement would be replacement blades for those still in the hands of private owners, as well as the preserved examples of the Battle of Britain Memorial Flight.

By this date, companies making wood blades for aircraft propellers were few and far between, although Permali Ltd of Gloucester, who had taken over the Hordern-Richmond firm, still made helicopter blades from similar material. Preliminary discussions with Permali began, but because the cost was considered too high, Dowty Rotol turned to the German company of Hoffmann, who already produced wooden propeller blades for light aircraft and eventually a sizeable order was placed for brand new blades. These are no longer made from the glued blocks already described, but instead use Birch veneers for the highly-stressed root area, combined with a normal density wood for the rest of the blade, this having similar qualities to the original spruce from which the Airscrew Company made propeller blades. The use of a lower density timber for the blade aerofoil necessitated increasing the finished blade thickness by some 10–15% to allow for stress considerations. However, owners of these old aeroplanes may rest assured that although the new propeller blades are not made to the original specification, they have passed all the required tests for airworthiness purposes and can be used with every confidence.

Readers may be interested to learn that a Rotol three-blade wooden propeller fitted to Spitfire IIa P7350 at Castle Bromwich in January 1940 is still flying at the time of writing. It first flew with 266 and 603 squadrons in 1940 and also served with 64 and 616 squadrons before being assigned to training duties and was preserved after the war. It took part in *The Battle of Britain* film and has been fitted with a Merlin engine from a Spitfire XVI but still retains the Rotol airscrew, and has now been flying for 50 years.

Into the 'Fifties

The 1950's were a time of great turbulence for the aviation industry, and this was inevitably reflected at Rotol. The decade began with doubts as to the direction in which the company should be heading, and ended with it being part of the Dowty Group of Companies. Between these two dates a great deal was achieved but the company altered considerably in the process.

In 1951, prompted by the outbreak of the Korean War, a further expansion of both works and offices became necessary. The 'White Block' received another storey, bringing it to its present height while the huts recently vacated by British Messier near the gatehouse were taken down and replaced with a brand new structure. This building was used as a machine shop at the outset, but after a while it was converted into offices, in which form it has remained ever since. Exactly why it should have received the name 'Stalag' – which was the name for German Prisoner of War camps – is not certain, as it does not appear to have any more sinister overtones than many of the others on the site.

The first event of note in the decade, was the appointment of Lt Gen Sir John Evetts as the first Managing Director of Rotol and British Messier in 1951. He took over from John Reid, ending the untidy arrangement which had prevailed until that time, when a General Manager had charge of the day-to-day running of the company, but without a seat on the Board of Directors.

Born in 1891, Sir John had a long and distinguished army career in both world wars, until he retired in 1946. In the First World War, he was wounded on the Western Front, mentioned in dispatches and awarded the Military Cross. He fought at the Battle of the Somme in 1916, as a major in the Machine Gun Corps, at a time when the life expectancy of front-line officers could be measured in days.

Between the wars he served mainly in the Middle East, becoming fluent in Arabic. When the Second World War began, he became Brigadier to the General Staff at HQ Northern Command India, but by 1941 he was in Egypt with the 6th Infantry Division, engaged in operations against the Germans and pro-Vichy French forces in Syria, who had established bases there. Once more, Evetts was mentioned in dispatches. He returned to Britain in 1942 to take up the post of Assistant Chief of the Imperial General Staff and in 1944 became Senior Military Adviser to the Ministry of Supply, where he was closely concerned with arranging the provision of a wide variety of weapons for the Normandy invasion.

He spent the years from 1946 to 1951 as head of the British Ministry of Supply staff in Australia, where he took a leading part in guided missile production and testing at the Woomera Range. An airfield at the range was named Evetts Field by the Australians, in recognition of his services. He was appointed CBE in 1937 and CB in 1939, receiving his knighthood in 1951.

Thus, when Sir John arrived at Rotol he already had a long and distinguished military career behind him, and at sixty years of age he began a second career in industry. How successful was he in this position? His knowledge of industry was not great and he obviously relied heavily on his executive staff, the majority of whom had served the company long and faithfully for many years. However, Sir John made up for any possible shortcomings with a great deal of experience of handling personnel, gained through a lifetime in the British Army.

Although of rather stern appearance, he was a highly-respected figure at Rotol and everyone who knew Sir John Evetts spoke of him as a perfect gentleman. When he walked through the machine shops, he knew many of the workers and shop stewards by name, and often found time for a friendly word, or to ask after a member of the family, who might be unwell. On occasions, Sir John was known to stop and offer a lift in the company car to employees waiting at the bus stop on the main road and it was his custom at Christmas to walk through all the offices and wish the employees the compliments of the season. Towards the end of his long career he devoted much of his time to Civil Defence, with particular emphasis on the threat of nuclear warfare.

The early years of the decade for the aviation world were dominated in Britain by the vast programme of turbojet and turboprop development being undertaken. Numerous 'flying test-beds' were engaged in proving new power units which the engine companies were busily turning out to suit a variety of requirements. Rotol was well represented on a number of these and on many of the aircraft participating in the Farnborough Display for 1951. These included the Bristol Brigand, Westland Wyvern, Fairey Firefly, Fairey Gannet, Vickers Varsity and Valetta, Percival Provost, the Vickers Viscount, and the last piston-engined fighter the Hawker Sea Fury. The star of the show was undoubtedly the giant Brabazon, with its Rotol contraprops.

Arguably the most important propeller ever to come through the machine shops at Rotol in the post war years, was that associated with the Rolls-Royce Dart turbine engine. It has been one of the company's most profitable ventures for a great many years and the supply of spares seem likely to continue well into the next century. Because of its impact on aviation, it is necessary to devote a few paragraphs to a short history of this engine and its propeller. The Dart engine was fitted to a whole host of aircraft in the fifties and sixties, almost all with Rotol propellers, but it was with the Vickers Viscount that the Dart will always be mainly associated.

With the advent of the turbine engine, a considerable discussion ensued within the aviation industry concerning the merits of the pure jet engine as against the turbo-prop and in order to rationalise the various theoretical arguments, Dr Stanley Hooker who was Chief Development Engineer at Rolls-Royce, suggested that it would be possible to modify an existing engine to give some practical answers. This idea was accepted and while Rolls-Royce began design work on a jet engine with a propeller shaft incorporating a reduction gear which eventually became the Trent, they also modified one of their early Derwent jet engines in order to save time. This ran for the first time in March 1945, fitted with a five-blade Rotol propeller having a diameter of 7′11″. Rolls-Royce followed this by fitting a pair of Trents to a suitably modified early Meteor F.1

EE227 and the aircraft made its first flight on 20th September 1945. The propeller diameter was gradually cropped to a little under 5 ft in a series of subsequent trials, providing some valuable practical data for the aerodynamicists. In the next three years, almost all the well-known test pilots of the period flew EE227 in order to gain experience with turboprop engines.

The Brabazon Committee had been set-up to look after British aviation in the post-war era and they were considering a replacement for the Viking, the first post-war airliner. Rex Pierson, then Vickers Chief Designer, put before the committee the proposal for a short to medium range aircraft, powered by four propeller turbine engines. This project culminated in Vickers Project Type No 453 and work began on two prototypes in December 1946. The Ministry of Supply were funding the development costs (although in the beginning it had started off as a private venture on the part of Vickers) and both Rotol and De Havilland were asked to produce suitable propellers. After a time, the Ministry realised that the double costs was wasteful to the British tax-payer, and a meeting was held to reconcile the position. Because Rotol and Rolls-Royce had already made a certain amount of progess with the Dart, it was agreed that Rotol would confine their turbo-propeller development to engine power ranges up to 2,000 bhp and De Havilland would cater for those above that figure. This arrangement gave Rotol propeller design authority for the Rolls-Royce Dart and the Armstrong-Siddeley Mamba (in its single and double versions). De Havilland had the Rolls-Royce Tweed and Tyne engines and the Bristol Proteus, this latter engine being chosen for the Bristol Britannia.

George Edwards (later Sir George) had taken over the post of Chief Designer at Vickers, when Pierson was made Chief Engineer, and despite the fact that he considered the Rolls-Royce Dart the obvious choice, the Ministry of Supply insisted on the Armstrong-Siddeley Mamba for the power plant. Within a few months this decision was reversed and test flying of the Dart went ahead in 1947, in a modified Lancaster and Wellington. Work on two prototypes to the finalised design, originally to be named Viceroy but changed to Viscount because of the partition of India in 1947, commenced in December 1946. On July 16th 1948 the prototype made its maiden flight, following this with an appearance at the SBAC Display in September. The aircraft was put into passenger operation by British European Airways at the end of July 1950, the world's first turbine- engined civil aeroplane.

Arthur Hayward was a technician in the Research Department at Rotol and was closely involved with the Viscount during its early trials. He recalled being in the aircraft when it was flown in icing conditions over Scotland, and due to the formation of ice in the intakes, first one engine and then another suffered 'flame-out', to be followed by a third. Tension in the cabin was mounting and the Vickers Test Pilot, Jock Bryce, turned to the Rolls-Royce engineer Frank Poole and asked, 'What happens now?' Over the intercom Frank gave the terse reply, 'Expect the last one to cut out at any time.' Fortune smiled on them that day, as the fourth engine kept going and Jock managed to restart the others. Trials continued with the Viscount in Hudson Bay, Canada, where it was left outside in –40°C. Let Arthur tell the story:

> We arrived in the afternoon at the Canadian Air Force base and parked the Viscount outside the hangar. We then went into a large well-insulated wooden structure, where there were numerous rooms leading off a central corridor, and were shown to our living quarters. In the morning we donned our special winter clothing before venturing into the outdoors but the cold was so intense that my nose froze up in the first thirty yards! When I reached the aircraft I found considerable chaos and an almighty mess, oil dripping from the engine cowlings and the snow turned into a quagmire of oil and slush. Ron Baker, an ex-bush pilot was Trans Canada Airlines Chief Pilot and he was standing on a trestle, pumping oil into the engine from a tank on the ground. I looked up at him and casually remarked, 'I see that we've got a bit of a weep, Ron'. He took a moment or two before he replied, in his broad Canadian accent – 'Yes Arthur, WE do seem to have a bit of a weep'. The Rotol oil seals in the propeller had not been able to cope with the extreme cold and an urgent message was sent off to the factory to get some that would. We were stranded for ten days while waiting for them to arrive, during which time I learnt some of the tricks of survival from the Eskimos, such as how to build an igloo.

In the next two years, orders began to flow in from other airlines, culminating in a 'first' from the North American continent, with an order from Trans Canada Airlines for 15 aircraft. This was followed by an order from Capital Airlines in the USA which they increased in the years between 1954 and 1958, making an eventual total of 60. By 1956, more Viscounts were flying in the United States and Canada than in the whole of Europe. This was not only a tremendous breakthrough for Britain but a considerable blow to the American aviation industry, which up to that time had a vast share not only of their own market but also that of the rest of the world. In an attempt to stop the importation of Rotol propellers on the Viscount the Hamilton Standard Propeller Company invoked the US anti-trust laws. Broadly speaking, these laws prohibited the import of foreign equipment if it could be shown that such imports would create a loss of business for American labour. A lengthy legal wrangle ensued which resulted in Hamilton withdrawing their objections, possibly because they realised that the time scale would prevent them from offering a suitable replacement, only to come back with an accusation that Rotol had infringed twelve of their patents. Another round of argument began, during which time LGF had to make several flights to New York for consultation with an attorney and a very capable patents engineer who had been loaned by the Bristol Aeroplane Company. Eventually, after months of discussion, the claim was finally whittled down to a single patent for which Rotol made a nominal payment.

Readers may be interested to note that by 1959, 400 Viscounts (with 1,600 Rotol propellers) were in service and Dart engines had amassed more than ten million engine hours in airline service. Over the years, the Rolls-Royce Dart was chosen to power many other aeroplanes, one of the most successful being the Fokker Friendship and its Fairchild licence-built version in the USA. At the time of writing, the total Dart engine hours must be approaching 100 million.

The Rotol management had been looking for some time into the possibility of creating a special facility in the United States, to cater for the growing number of Rotol products that were in use by American and Canadian airlines. Product support, for both the propellers and the accessory gearboxes fitted to Capital

Airlines Viscounts, was provided initially by Vickers out of their Arlington facility in Virginia. This was not particularly satisfactory, so Rotol decided to set up their own organisation. The result was the formation of Rotol Incorporated on September 18th 1957, which commenced trading from a corner of Vicker's warehouse on Jefferson Davis Highway. Bryan Clark, Rotol Sales Manager, was instrumental in obtaining the services of Vernon Crudge, who had a very good standing with the top executive personnel throughout the American aircraft industry, having operated a consultative business in New York for some years, and he was appointed a director of the new company. The President of Rotol Inc was Jack Staples who had been Service Manager at Rotol for some years, and Howard Bird was recruited as Treasurer. Existing service engineers stationed in the United States were transferred to the payroll and the new organisation quickly began to make a name for itself. Mike Brown, one of the draughtsmen from the Main DO was given the post of Office Manager but was tragically murdered after a few short months.

Mike will be best remembered by the draughtsmen of the time as one of the office jokers, often in collaboration with Wilf Gordon. The two of them were working on overtime one evening when they came up with an idea which was to subsequently cause considerable discomfort to the rest of the Drawing Office staff. Bill Parry, who ran the Standards Section, was renowned as a rather politically argumentative individual and Mike and Wilf decided to get their own back on Bill by unscrewing one of the legs on his metal table and placing a kipper inside one of its hollow legs. Within a few days the smell in the office had become indescribable and it was some time before its source was located. On another occasion, when the new adhesive 'Araldite' first became available in the workshops, they glued a sixpenny piece to the floor and the rest of the office watched in delight as a succession of higher management figures attempted to pick it up. Glueing coins to the floor may be 'old hat' today, but this was almost certainly the first occasion that the trick was tried at Rotol. Both Mike and Wilf have long since passed away, but they live on in memory.

Back in the United States the repair and overhaul facility catering for the Rotol and British Messier products became so successful that five separate sets of premises had to be found around Arlington. The equipment ranged over a wide variety of aircraft, including the Dart propellers and gearboxes on the Viscount, Fairchild F27 and Grumman Gulfstream, and the undercarriages and hydraulics for the Bristol Britannia, all of which were amassing a great many flying hours, both in the United States and other places overseas. Having so many separate sites was not conducive to efficient working and a search was made for a larger factory. This was eventually found on a virgin 40 acre site at Dulles, Washington and since that time the facility has doubled in size, providing a spares depot and a repair and overhaul base for all Dowty Group products, as well as those of Boeing and the French Messier Company.

This long excursion into the tale of the Rolls-Royce Dart and the Viscount, and the resulting developments, has taken us away from the other events of the decade. We must return to the beginning of the 1950's to pick up the unfolding history of the company.

Work on reversing propellers, which had started during the war, was still under development by Rotol, as their use would be helpful as a braking aid on landing. Test flying of a Lancaster fitted with normal propellers to the two outboard engines and Rotol braking types to the two inners, was continuing at Moreton Valence. On touchdown, reverse pitch operation would be selected and the two inner engines would be opened up in the braking mode to slow the aircraft – at least that was the theory – but on Friday June 9th 1950, something went wrong. As the aircraft came in to land, some 200 yards from the runway and at a height of perhaps 100 feet, the pilot, John Cree, opened the throttles slightly to adjust the approach and the inner propellers went into reverse. By some miracle, Cree managed to hold the nose level and the Lancaster bellied in. The fuselage broke in two but remained attached, the engines broke from their bearers and one wheel was forced through the wing. Thankfully, there was no fire, and the rescue squad were quickly on the scene. On board was the pilot, who suffered a fractured spine and arm, and four others, John Turner (who later took charge of the Dowty Rotol Service Department and retired in 1990), 'Jimmy' James, (retired 1982) and Edward Kelly and John ('Joe') Mitchinson, all of whom received treatment in the Gloucester Royal Infirmary. The cause of the crash was never conclusively established, and despite exhaustive tests the reversing propellers could not be made to malfunction.

Following the Lancaster enquiry it was decided that because Moreton Valence was separated from the Rotol Design Office by 14 miles, the FTD warranted a permanent representative from the DO and Jack Meaden was probably an obvious choice, with his piloting knowledge and technical background. A drawing board and desk were set up in Ben Harris's office and Jack settled down to a job which was 'right up his street'.

When the Flight Test Department closed down in 1954, Jack Meaden returned to Main Works for an interview with George Bubb, and was told that LGF had decided that he should work on the merging of Rotol and British Messier design standards, which was something of a thankless task as most changes had to be implemented 'by persuasion'. He joined the Standards Section of which W.E. (Bill) Parry was the Section Leader. Bill's support for Communism was well known and he frequently got into heated political arguments with others in the Drawing Office.

The Rotol Design Standards Book, issued to all members of the Main Drawing Office and Project Design Office, had its beginnings as 'DO Standards' consisting of a number of memos issued by the Chief Draughtsman A.E. (Eric) Stockdale and a selection of national and British standards prepared by Bill Parry. In spite of their importance in everyday use, except for those few regarded as mandatory, the project designers were prone to treat them as limitations on their freedom to make their design work easier by pleasing themselves on design parameters, than as standards with the purpose of making interchangeability easier and reducing the cost of parts and tools to a minimum. This occasional negative attitude on the part of the designers particularly infuriated Jim Tyler, a member of the Standards Section, who would argue fiercely in support of working to the system.

Jack Meaden eventually took over the section when Bill Parry left. By then the services of Standards had expanded enormously to the extent that it supplied the whole of the company with standards and specifications, plus advice on their use. It was renamed Standards Office in recognition of this, but the old hands still thought of it as only a section of the Main DO and called it Standards Section from force of habit.

Rotol had established itself as one of the top propeller manufacturers in the world but it was always sound policy to have other strings to one's bow. Representatives of the Technical and Sales Departments were continually on the look-out for other aircraft products, which could be developed into profitable items, and the 1950's saw the introduction of several such items. The main responsibility for locating new projects rested in the Sales Department and it is there we will meet another well known employee, although I regret that I have not been able to unearth more of his background. Bert Newport came from BAC where he had been a technical representative for the Engine Division, working mainly in Italy. In 1937 he transferred to Rotol as one of the first employees, firstly as Development Engineer and in January 1940 was promoted to Assistant Chief Engineer, a post he was to hold until 1952. In that year he went to Australia to start a Rotol Agency office in Melbourne, returning after two years to become the Technical Sales Manager. When Bryan Sandford Clark, the Sales Manager left the company to take an appointment as European representative of the Garrett Corporation of USA he was succeeded by Bert Newport and remained in this position until he retired. Bert was summed up by one of his compatriots as having a reputation for being a first-rate engineer who always knew what he wanted.

The first of the new Rotol products was the Fuel Flow Proportioner, designed to supply the right amount of fuel from mutiple fuel tanks to the engine or engines, making it unnecessary for the pilot to monitor fuel except for total endurance. They were first introduced in 1950 and continued to be a useful and relatively simple piece of equipment for many years. Fuel Flow Proportioners were only fitted to military aircraft, the reason being that military pilots needed to be able to give their attention to the military operation on which they were engaged and could not spend time monitoring fuel as could the commercial pilots and flight engineers. With the introduction of the Rotol Fuel Flow Proportioner this function was carried out automatically. In essence, the proportioner is a series of vane-type positive displacement pumps, the rotors of which are mechanically coupled. The displacement of each pump 'cell' is proportional to the capacity of the tank it is intended to serve.

The production version of a twin-cell proportioner was fitted to the Hawker Hunter, used extensively by many air forces all over the world, to be followed by a multi-cell version, for the original Handley Page Victor. The Hawker-Siddeley Harrier, the revolutionary VTOL aircraft, also carries a Rotol Fuel Flow Proportioner. With only minor changes, this particular piece of equipment has been fitted to every variant of the aircraft produced to date, from the original Hawker P.1127 prototype in the late fifties to the Harrier GR.5 of 1989. Many other aircraft flew with a Rotol Proportioner, including the Italian Fiat G.91 fighter supplied in large numbers to NATO Air Forces in the 1960's.

The Rotol ram-air turbine, also introduced in the 1950's, was developed because of the need to provide a reliable and independent power source, should the main engine or accessory drive fail. This can be a serious event and in a single-engined aircraft disastrous, as in most high performance aircraft the flying controls are operated by hydraulic power supplied from the main engine.

The AiResearch Manufacturing Company of the Garrett Corporation of America, had evolved a small lightweight turbine which could be extended into the air stream. The turbine was coupled to a hydraulic pump connected to the aircraft hydraulic system, but a later development used on the Vickers VC.10 had an electrical generator instead. Rotol made these under licence to start with, but quickly began to design a whole range of variants of this most useful piece of equipment. Over the years, a number of aircraft owe their survival to the Rotol ram-air turbine.

Air Turbine Accessory Drive Units, (not to be confused with ram-air turbines) was another product which initially came from the AiResearch stables. The first of these utilised an AiResearch air turbine with a variable nozzle, coupled through a Rotol gearbox to a 15 kva alternator and a 3 kw generator. The turbine was driven by compressed air, tapped from the main gas-turbine engine compressor. The Air Turbine Unit was developed to overcome the basic disadvantage of the normal shaft-driven gearbox, the inability to operate accessories at constant-speed, essential for radar and other instruments, requiring a stable input frequency of alternating current. Production versions of this new equipment were fitted to both the English Electric Lightning, the first supersonic fighter to be used by the Royal Air Force, and also the Blackburn Buccaneer naval aircraft.

Ken Jones, a first-rate engineer with an intuitive feel for his job and anything connected with air turbine equipment, was the Turbine Test House Foreman at a time when the P.1 (the prototype Lightning) was undergoing its development period. Ivor Brooking, the designer responsible for the Lightning equipment, was also present on this particular occasion. It was nearly midnight and the test engineers had been attempting to clear the turbine for some hours, without any sign of success, when Ken turned to Tom Godden, the Chief Engineer, who was witnessing their struggles and said in exasperation, 'Tom, from where I'm standing, we have as much chance of getting this unit to govern correctly, as you have of adjusting the height of your bathwater by playing with the tap in the kitchen of the house next door.' Sadly, both Tom Godden and Ken Jones were to die at a tragically early age.

At the end of July 1953, some important changes were announced in the composition of the Rotol and British Messier Board of Directors. Lord Hives, the joint managing director of Rolls-Royce and a highly respected figure in British engineering and Sir G. Stanley White, deputy chairman of the Bristol Aeroplane Company, who had both been with the Rotol Board since the first days in May 1937, resigned from the board as from July 24th, and for the time being no one was nominated to fill the vacancies. The Rotol Board then consisted of the following: A.G. Elliott (joint managing director, Rolls-Royce), Sir W. Reginald Verdon Smith (joint managing director, Bristol Aeroplane Company), W.T. Gill (financial director, Rolls-Royce), Dr A.E. Russell (technical director, BAC) and Sir John Evetts (managing director). At the same meeting, Col C.W. King was appointed general works manager of Rotol from August 1st 1953. It was to be more than a year before further changes took place in the Rotol Board, when Air

Commodore F.R. Banks replaced Dr A.E. Russell at the end of 1954 and R.J. Raff took over the position of Company Secretary from D.A. Tomlinson.

The closure of the Rotol Flight Test Department in 1954 did not mean the end of propeller testing of course, and the auto-observers and other test equipment developed at Moreton Valence continued to be used by the Rotol Development Department. The testing of propellers and associated equipment became part of the maker's testing of prototype aircraft and this involved the Rotol test observers visiting each aircraft manufacturer's airfield, with their equipment vehicle.

In October 1954, Jack Meaden, who was now back at Rotol in the Main DO was asked by Eric Fairhurst if he would like to 'volunteer' to complete the testing on the Napier Eland Varsity, as no test observers were available. This would involve about a week's flying from Napier's Flight Test Unit at Luton, to which Jack agreed. The flights during the week were routine and uneventful until the Sunday, when the final flight on the programme, a diving test over the Atlantic off the French coast, took place. When the last dive was completed and the aircraft levelled off above a layer of low stratus cloud covering Southern England, Jack signed the last sheet of test figures, switched off the equipment and sat back to enjoy the rest of the flight. When he glanced out of the side window he saw a stream of oil pouring back from the port engine and over the nacelle. Jumping forward into the cockpit he alerted Ken Cartwright, the pilot, and sat in while Ken took a quick look. The pilot was now faced with a crucial decision. If he stopped the engine and feathered the propeller the aircraft at test weight might not be able to maintain height and reach the airfield on the one remaining engine – which was not giving the designed power. If the engine failed for lack of oil before he could feather the engine, the propeller would be in coarse pitch and the resulting offset drag would make the aircraft difficult to control, an emergency situation of which all twin-engine pilots were mindful. Jack sat alongside the pilot, ready to help on the controls if this happened but Ken elected to throttle back the leaking engine and began a slow descent through the cloud layer. After what seemed an age, but was possibly only a few minutes, they broke out of cloud and saw the Luton cut-grass runway lined up dead ahead. Ken cut the ailing engine as they crossed the boundary hedge and landed safely on the other. They were towed back to the parking area and the clamshell cowlings were opened, releasing several gallons of oil that had collected in the lower cowling, on to the ground. A check in the oil tank showed that they had only about two pints of the original eight gallons left, this being the margin by which Ken, Jack and a Napier apprentice who had come along for the ride, survived what looked like being a very desperate situation. The oil leak had been caused by an unlocked union nut that had come loose on one of the test equipment pipes, and there was no fault in the engine or propeller.

Lord Hives may have severed his official connection with Rotol but he was asked to perform the opening ceremony when a brand new Apprentice School was opened in September 1954. Like many of the departments at Rotol, the School had occupied a number of different sites. In the mid 1950's it was in the Gatehouse (known universally as 'Jelley's Folly') where the machines were on the ground floor and the offices upstairs. When it moved to a purpose-built workshop and offices adjacent to the gate into Hatherley Lane (replacing some wooden huts which had housed the labour offices) many must have thought that their moving days were over. The building was perfect for the Apprentice School, and very much appreciated by the 200 apprentices, but it moved several times over the succeeding years, on occasion to the South Works on the airfield side of the Gloucester–Cheltenham road, then back to sites in the North Works. There were undoubtedly good reasons for all these moves but whatever they may have been, they are lost in the mists of time!

Sam Peebles was in charge of the Apprentice School, greatly helped by the instructors – Charlie Cuthbert, 'Roly' Clutterbuck, Johnny Bunkell and Bill Green. One of the apprentices during the 1950's was Brian Watts whose story is told below. He recalled that although Johnny Bunkell was not a large man, he was quite strong, and when he tightened up a four-jaw chuck it was almost impossible for any of the boys to release it without some mechanical advantage! Bunkell was reputed to have webbed fingers which probably contributed to his powerful grip.

A good many ex-apprentices responded to my request for information, and almost everyone had a story to tell concerning their early days with the instructors in the Apprentice School. These stories would probably fill a book in themselves. The natural high spirits of the boys was always a source of some mischievous plot against either higher authority or one of their own number, and there are few who will not have been party to a mild form of mayhem during their apprenticeship days.

Initial training was in the school and the lads would then go off into the factory, or 'On the job training' as it is now described. Few girls have ever undertaken an engineering apprenticeship, with some notable exceptions in more recent years, but in the early days it was unheard of. In the fifties, apprentices working in the Production Machine Shops were on the same bonus rate as the rest of the machinists, but after a time the arrangement was cancelled and subsequently if an apprentice worked on an opposite shift to a production worker, his bonus was increased by the effort of the young lad on the other shift. Needless to say, as Brian Watts has confirmed, this new system was not welcomed. Brian contributed a tale of the time when he was an apprentice in 1956 and had served his time in the school. Now he was working as part of the assembly test team, in what was then the Airframe Development Department:

> The hydraulic testing had to be approved by an inspector who on this occasion was Bob Hutchings. Bob was noted as a man of high personal cleanliness, with a careful regard for his appearance. He always dusted a seat before he sat down, and it became something of a crusade by the rest of the men in the department, to try to get Bob as dirty as possible. Despite all their scheming, Bob led a seemingly charmed life. No matter what traps were laid for him, none were successful. On this particular occasion, Brian carefully set a thermal relief valve so that when it 'cracked off' the resulting spurt of oil would hit the ceiling and spread a film of oil over anyone close by. The valve passed all the rehearsals with flying colours and Bob was called over to check its performance. While his mates watched with unconcealed delight, Brian pumped away on the hand pump and Bob bent over to watch more closely, and everyone tensed as the appropriate pressure drew near. The valve gave a very gentle crack, well within the set limits and Bob asked for a repeat, just to be sure. Once

> again it worked perfectly and Bob happily signed the sheet and walked off. In disgust, Brian did another run and got soaked from the resultant blow-off – once more Bob Hutchings had got away with it!

Before we leave the Airframe Development Hangar we can have a smile over a story from George Brazil, who was working there on night shift. The charge hand was very proud of his small Austin car and some of the lads got together in the car park and lifted it on to four oil drums during the dinner break. The expression on his face when the shift finished and he strolled out into the morning sunshine, was well worth seeing!

The role of the trade unions at Rotol has not so far been given much of an airing. Although industrial relations is an important topic, there are few stories which the general reader will find of interest but the trade union men and women who represented their fellow members will possibly recall C.N. Taylor, a well-known figure for many years. Cyril Taylor had been with the Glevum toy factory for eight years, prior to taking employment with Rotol as the Chief Industrial Welfare Officer during the war and in later years he was appointed to the position of Personnel Manager.

Ron Pirie, a designer in the Project Office, played a major part at Rotol in the AESD (Association of Enginering and Shipbuilding Draughtsmen) and related the following tale just two weeks before his death:

> Taylor had a rather unfortunate tendency to appear somewhat aloof when listening to a union deputation. In addition he had the habit of paring his finger nails with a penknife during the meeting, much to the annoyance of all concerned. One day in my role as union representative from the Drawing Office, I decided I could stand it no longer and the following weekend I purchased the largest nail file I could find in Cheltenham. At the next confrontation, Taylor pulled out his penknife, and I quickly followed with my nail file. Taylor put down the knife and the discussion continued, but whenever the penknife re-appeared, I responded with my file. The message was clear, and from that moment onward, Taylor always paid far more attention to the matter in hand.

One of the main characters on the technical side who has been mentioned only briefly was Ellis Danvers – although no one ever referred to him as anything other than 'Dan', as he had no liking for his given name. Ron Pirie retold the story of his first encounter with Dan, who was then Chief Designer, shortly after he arrived at Rotol as a Project Designer. Ron's first job, under the eye of Arthur Lane his section leader, was the preparation of a gearbox scheme for the Sea Vixen and after some weeks intensive work it was finally completed. Before Arthur could bring Dan to view the result and give it his approval Ron decided that it was so 'tatty' after weeks of alteration that the five sheets needed redrawing. In a concentrated effort over four days he accomplished this task and proudly awaited the arrival of Dan. When he turned up he contemptuously dismissed Ron's best efforts with the curt comment 'Not enough rubbing out', and marched off. After allowing some time to elapse Ron produced the 'old' scheme, complete with all the alterations, erasing etc. and Dan accepted it without a murmur. It was Ron Pirie who reminded the author of one of Dan's favourite sayings, 'The man who never made a mistake never made anything.'

It was sometime in the 1950's that another story originated which was destined to be repeated on many occasions. A brand-new propeller was urgently required for testing at the Royal Aircraft Establishment at Farnborough. Because it was somewhat larger than previous propellers, considerable attention was paid to the route which the lorry would take in order to avoid any low bridges. On the day in question the propeller was loaded and secured to the lorry with great care, and the necessary documents were doubly-scrutinised to ensure they were all in order before the lorry driver set off. The vehicle had started to move, when somone shouted a warning, just in time to prevent the propeller being severely damaged. Despite all the care which had been taken in planning the route, nobody had considered the overhead steam pipes, which crossed the road just before the factory gate. The propeller was off-loaded and put back on the lorry, on the other side of the offending pipes, and sent safely on its way.

Fred Holtham spent his years with Rotol as a driver, and his tale concerning propellers did not have such a happy ending. It was the time of the Python-engined Wyvern, and Fred was taking his customary delivery route to the Westland Aircraft Company at Yeovil with a lorry loaded with the large Wyvern contra-props. He was making good progress, when he came to a bridge carrying a railway, which had not been a problem in the past, but what Fred did not know was that the road had been 'improved' and the surface was now somewhat higher! He was soon made aware of the changes as the propellers struck the underside of the arch, with the result that several of them were knocked off the lorry and into the road. His explanation on his eventual return to Rotol must have been accepted, as Fred continued to drive for the company until his retirement in 1979.

Stories concerning the people who worked in the Experimental Department are legion, possibly because it had more than its fair share of characters, coupled with a loyalty and a reputation which it had to uphold. As an example, Arthur Gunn told me of one of the machinists who was taken ill with tubercolosis and the rest of his mates collected together a weekly food parcel for his young family, which they continued to supply until his recovery and eventual return to Rotol.

The Department was noted for being in the forefront of everything from wage demands to practical jokes, as the following tale related by Eric Dyer will show. Eric started at Rotol in October 1940, first in No 1 Machine Shop, transferring after six months to the Experimental Department, where he worked as a centre lathe turner right through the war. In 1947 he took a staff appointment in the Experimental Planning Section and, as a keen cricketer, Eric played both for the department and the Rotol First XI. In the fifties, when the department was under some very heavy pressure to complete a job, Les Mercer who worked alongside Eric, came up with an idea to demonstrate just how busy they were. When the telephone next rang, Les answered by telling the caller to ring the new number, which had just been installed in the toilets in an effort to cut down on the waiting time. This number, of course, was that belonging to Eric, who picked up the phone and announced, 'Gent's toilets'. This unexpected reply dumbfounded the caller and all he could say was, 'Well I'm blowed'! (These are not the precise words which were used; the reader can probably imagine that the caller was somewhat more forthright.)

Frank Loveridge spent well over thirty years in the Experimental Department, joining in 1952, first as a fitter, then an inspector and finally as a Methods Engineer in the Experimental Office. He had a fund of stories to relate, mainly concerning disasters in the test house where he worked for a period, involving the personnel diving for cover when something broke, but he admits to almost being sacked for misconduct on one occasion. He was building a governor unit in the Assembly Shop and as the build progressed he took the parts from the carboard boxes, stacking the discarded boxes on the next bench. Roy Mills was working on a nearby bench and feeling rather mischievous, began to toss his empty boxes on to Frank's bench. It was only a matter of time before Frank tired of this and hurled one of the boxes back, straight through a glass partition. This in itself would not have been quite so bad if the senior foreman, Ted Lethbridge, had not been on the other side, conducting a party of important visitors round the department. Fortunately for Frank, he had no more than a reprimand for his foolish action and the matter was soon forgotten.

A familiar face came back to the company, during the mid-fifties, after a break of several years. Cyril J. Luby, the first Rotol Works Manager, had left in October 1942 to rejoin the Bristol Aeroplane Company but returned in 1956 in the capacity of Assistant Managing Director. Other changes, which took place around the same time, were the appointment of a further three directors to the Rotol Board. The Rolls-Royce Executive Vice-Chairman, Whitney Straight, took A.G. Elliott's place on his retirement, and Col C. W. King moved from Works Manager to Works Director. In recognition of his long service to the company, L.G. Fairhurst was promoted from the position of Chief Engineer to become Rotol's first Technical Director.

In the mid-fifties, although Rotol were heavily engaged on the development of the Dart propeller, they also devoted much time and effort on a complex system of twin turbo-propellers and associated equipment for the Fairey Rotodyne, a revolutionary passenger helicopter powered by two Napier Eland engines. There were some doubts concerning the commercial future of this aircaft and a few months later these doubts became reality when the project was cancelled.

The fact must not be overlooked that Rotol equipment on Dart-powered aircraft was not confined to the propellers alone. In addition there were governor units, feathering pumps, synchronisers and gearboxes and this is perhaps an appropriate moment to mention the electrical side of the Dart propeller equipment, as it made a great contribution to the success of the Rotol product. The Electrical Design and Development came under the overall control of L.G. Fairhurst as Chief Engineer, and Alf Mapp the Chief Electrical Engineer directly resonsible for all the day-to-day running of this part of the organisation, with rather less than a dozen technical assistants. John Cripps was Alf's assistant, to whom I am indebted for loan of material on the subject of electrics at Rotol appertaining to the late 1950's. John took over from Donald McKenzie, who is believed to have disappeared in one of the redundancies of the period. The Electrical Design DO came under Sam Warriner, who had charge of a small number of electrical designers and draughtsmen, among whom were Jack Beach, Tony Evans, Dick Williams and Dave Soley.

Electrical de-icing of the propeller blades was the subject of a long and painstaking development over many years, culminating eventually in the heater elements fitted to the Dart propeller blades and spinners. Arthur Hayward had been closely associated with experiments in electrical de-icing since the very early days, when the Hydulignum blades had wires embedded in the wood surface. Flight testing was carried out on a Whitley bomber with radial alternators built into the rear of the airscrew, but the strain was too much for the small element wire and one of the airscrews equipped with this device failed even before the aircraft could take off. Eventually, further airborne tests took place in icing conditions when water was sprayed from special equipment onto the rotating blades until a sufficient quantity of ice had built up. The engine was then stopped, photographed and then restarted with a current applied until the ice melted. Charlie Harmer, a member of the Flight Test Department, was present when they were flying on one particularly warm day of high humidity – chosen because icing can then be achieved at lower altitudes – when a large block of ice began to form on the air intake grill of the Merlin engine, eventually breaking away and hurtling to earth in the vicinity of Charlton Kings. No reports of extra large hailstones appeared in the local papers, but the crew speculated on what the owner might think if he came home to find a large hole in the ceiling and a damp patch on the bedroom floor!

The use of metal spray applied to the blades to carry an electrical current was also considered and after a lengthy period of time a rig was devised by Reg Alexander to enable the spraying to be evenly distributed. Following protracted trials Rotol abandoned this method, mainly because of difficulties in getting a consistent thickness throughout its length. Any high spots had to be located and rubbed down by hand, in order to obtain a specific resistance, and although some companies are believed to still use this method, it was not considered satisfactory. The problem was aggravated by the Rotoloid which covered the wood blade, which had the tendency to ignite almost without warning and Arthur considered it a point of honour when the blades were undergoing test and the element heated up, to extinguish the resulting blaze before the arrival of the Rotol Fire Brigade.

Rotol in the meantime began to look into the use of wire-woven heating elements, based on the work done during the war on heated flying suits and gloves, but initial efforts were frustrated by too coarse a weave. Chief Project Designer, John Chilman had been closely involved in the design of electric de-icing. He and Tom Godden agreed that the solution lay in some form of element woven into a suitable material, through which a current could pass. Tom's wife had an interest in weaving, and a small loom, so she was asked to produce a simple element using material supplied by Chilman. This was the start of what was to become a most successful venture, to the extent that Mrs Godden even set up her own factory in Cheltenham, turning out heater elements for all manner of industrial uses, and although the original factory has been taken-over by a different manufacturer, heater elements are still being made at a small factory in nearby Charlton Kings.

One of the features of industrial life in the 1950's was the threat of redundancy, which might be caused by lack of orders, take-overs by rival companies or any other reason which meant that the workforce had to be reduced. A few weeks after Cyril Luby's appointment, he had the unenviable task of allaying fears of redundancy. Rotol had remained remarkably free of this particular spectre until now but the question was raised – for how long? Rumours had become rife throughout the factory and in a meeting of departmental heads and other executive staff he gave assurances that there was a full order book, with sufficient work for the next two years.

Despite this, rumours continued to circulate throughout the summer and the chill winds of the mid-fifties gave further cause for concern at the board meeting held on October 4th 1956. In the light of the fact that new orders were not forthcoming to the same degree as previously the Managing Director, Sir John Evetts, raised the possibility of the company looking beyond the aviation field for the future and setting up a separate industrial organisation, in parallel with the aircraft side. It was also stated at the same meeting that unless the situation improved, all activities would have to be concentrated in the North Works and the South Works closed down.

The chairman, Sir Reginald Verdon-Smith, summed up the views of the Board by urging that Rotol should endeavour to obtain a larger share of the accessory market, despite the fact that the market was tending to diminish. They should also try to find suitable selected products in the industrial field, and he ended with the wry comment that some forecasts made by companies in the aircraft business had proved wrong in the past.

H.L. ('Pop') Milner died on 19th December 1956. He was born in 1886 and his early career has been narrated in a previous chapter devoted to the history of variable pitch airscrews prior to the emergence of Rotol, but there are some stories concerning him which deserve recording. Apparently he was something of a joker and also played chess with Lt Col Blyth on occasions. One Christmas he gave Blyth a magnificent leather-bound book entitled *All I know about chess* – in which all the pages were blank! He could also appreciate it when the roles were reversed one dark night, when he had spent some futile minutes attempting to crank start his car and complaining bitterly. One of his colleagues passing by resolved the problem by pointing out that although the car was the right make and model it was not Pop's car which was situated a short distance away.

He was to make a further vitally important contribution to the propeller world which deserves recording. The permitted manufacturing tolerances of the finished propeller were often the cause of undesirable vibration in the aircraft, affecting the instruments and contributing to pilot fatigue, and during the war Milner began to look into ways and means of solving this problem. Dynamic balancing had been part of the manufacturing process since early days and aerodynamic balancing of propellers was known to be the answer, but achieving it was not so easy. By using caliper gauges and a clinometer it was possible to calibrate blades against a master blade, and although it was used in the early part of the war, it was far too time consuming, requiring skilled operators to make the routine calculations needed for each calibration.

The outcome of Milner's work, patented in 1941, was an apparatus which overcame all the basic disadvantages, permitting the overall aerodynamic correction for a propeller blade to be measured against a master blade and indicated on a graduated scale. A complete description of the workings of the blade integrating machine as it was called, involving a series of linkages and pulleys which at first sight might be attributed to Heath-Robinson, is rather too technical for inclusion here, but anyone who is sufficiently interested is referred to the paper which

Milner wrote for publication by the Royal Aeronautical Society in December 1949. In 1944, Milner and J.L. Owner, the Rotol Patents Officer, journeyed to the United States for discussions with interested propeller companies, this eventually resulting in the Rotol method of balancing being adopted by most of the industry, both in Britain and America and making a valuable contribution to the Allied victory. With the advent of the turbo-propeller, and the low level of vibration in the turbine engine, getting rid of anything in the propeller that contributed to vibration assumed an even greater importance.

'Pop' was appointed Project Engineer towards the end of his career, studying future trends in propeller and accessory design. He was always pleased to help youngsters with their technical problems and on one occasion he was in the throes of giving his assistance to an apprentice when the Managing Director rang to ask to see him. 'Pop' replied that he would be pleased to attend in a few minutes time when he had completed the job in hand and he then replaced the telephone and returned to the homework that was giving so much trouble.

Although his later years at Rotol were perhaps spent in a relative backwater as others made progress with his initial ideas, he had the satisfaction of knowing that in his lifetime he became recognised as one of the architects of the modern propeller. At his funeral he had the unique distinction of being carried by a bearer party of five Rotol departmental heads and one ex-apprentice – Roger Armstrong.

The Dowty Take-over

As the decade of the fifties drew to a close, every effort continued to be made to find markets for existing Rotol products and to develop new ones. The demand for propellers had been inevitably reduced with the upsurgence of the jet engine, which now powered almost all long-haul flights and gradually expanded into the holiday travel market. The Rolls-Royce Dart engine fitted with Rotol propellers, had a healthy share of both the short and medium range aircraft market at this time but it was always a sound policy to be on the look-out for future prospects. Although Rotol and British Messier were represented on the Britannia airliner, by virtue of the latters contract to produce the landing gear and hydraulics, the propellers were a De Havilland design.

Military equipment was no longer in such demand, due in part to the White Paper on Britain's future defence policy which had been published in April 1957. The Defence Minister in the Tory Government was Duncan Sandys and his proposal was to phase out development of manned fighter aircraft and replace them with missiles, coupled with a reduction in the number of bombers. This sent a shock wave throughout the British aviation industry and resulted in the cancellation of a number of promising aircraft. Several projects in the Rotol and British Messier Design Departments were affected and a pall of gloom was cast throughout the works. One of the only bright spots came in the amusing story that was told around the factory and offices concerning a cleaner in the toilets who was heard to remark that he could no longer see a future in aviation and was going to hand in his notice.

A pleasant surprise was introduced into the Apprentices' Prize Giving ceremony on May 17th 1957, when Sir John Evetts paid tribute to the Twentieth Anniversary of the company and invited seven members of the audience to step forward and receive a gold watch in recognition of their long service since that date. They were L.G. Fairhurst, H. Newport, R. Barrell, W. Brierley, C.H. Smith, J.H.M. Parker and R.H. James and in the next two years over 200 employees were to receive a similar token for their services. Sir John usually made it his practice to deliver the gold watch to each recipient at their workplace but the increasing numbers in later years made this impossible.

To commemorate the Twentieth Birthday the company held its first Open Day on July 13th 1957, when nearly 10,000 employees and their families looked around the workshops and offices. Later in the day they attended the 16th Annual Sports on the firm's athletic ground and watched the inter-departmental tug-of-war and relay race, won by the Experimental Heavy-Weights and the Main Drawing Office respectively, as well as all the other many attractions. These included a horticultural and handicraft show in a large marquee adjoining the sports field, and a traditional fair complete with side-shows and a Punch and

Judy show, and the day was rounded off with old time and modern dancing in the canteen.

It was during one of their periodic visits to the Garrett Corporation in America that LGF and Alan Walker – who had taken over the post of Chief Aerodynamicist from Ken Thatcher – were asked if Rotol would be interested in entering the field of turbo-charging of diesel engines for the industrial market. After some discussion, when the Rotol representatives were given every access to the work done by Garrett's thus far and the possibilties which seemed most promising, a decision was made to produce several prototypes. At the same time, with all the many personal contacts that LGF already had in the commercial vehicle industry, he followed this up with discussions with leading UK engine companies such as Leyland, Gardners, Perkins etc but although the value of turbo-charging was fully appreciated, in every case they warned Rotol of the danger from merely fitting a turbo-charger direct to an existing engine. They were adamant that only when the full assessment of the effects on their engines were known and a new range suitably modified to withstand the high stresses imposed, would they offer turbo-charging as an optional fitment.

The reactions from the engine companies gave Rotol the confidence to go ahead and on August 1st 1957, C.J. Luby, the Assistant Managing Director, was directed to set up a new Division to develop and manufacture turbo-chargers for the Diesel engine industry on a production basis. W.G. (Bill) Savory, a Project Designer for a number of years and originally responsible for the initial turbo-charger designs, was chosen to take charge of the new enterprise. The Turbo-Charger Division was an entirely separate organisation from the rest of the works and had its own machine shop, recently the pride and joy of the Apprentice School. The title was changed to Industrial Division and continued to operate at Rotol until the Dowty take-over, when it was transferred to Arle Court and became part of Dowty Hydraulic Units.

Despite the recent excursion into the industrial field, the question of the lack of new aircraft products was once more the main item for discussion by the Rotol Board of Directors at their meeting on October 3rd 1957, when it was suggested that from 1960 onwards, the company might have to consider producing even more equipment outside the aircraft field, or face up to reducing personnel and plant. After some considerable discussion had taken place concerning the general uncertainty felt by the Directors, a statement was made by W.T. Gill, 'that the Nominated Directors, (those from Rolls-Royce and BAC – the others were known as the Resident Directors) should meet at a convenient date, to discuss the future policy of the company.'

If it is possible to fix a point in time marking the beginning of the discussions which were to lead to the take-over by Dowty, I suggest that this was the moment. During the course of researching the history of Rotol, one question more than any other has cropped up in conversations with the author, 'What was the true reason for the company being sold to the Dowty Group?' No doubt much went on behind the scenes which has not been recorded, but from reference to the minutes of the Board of Directors' meetings one sees an unfolding series of events which pointed to only one conclusion that the Board could make.

R.J. Raff, the Rotol Secretary, gave warning at a meeting a few weeks later of the grave financial position which would develop in the early months of 1958, when there might be a risk of the company over-trading in relation to its financial resources. One of the reasons why this state of affairs had come about, was the policy of accepting certain orders on which the profit margins were lower than normal. This was quite usual in the aircraft field, where companies cut their tenders to the bone in an endeavour to get their equipment on new aircraft. Only a few months previously it had been agreed that selling prices for the early Dart propeller equipment would not be increased in 1958 or 1959, regardless of whether or not any National Wage Awards were made in those years. Possibly another factor was the large stocks-in-hand that the company possessed, which, although they would be converted into cash at some future date, did not solve the immediate financial problem.

Amidst all the rather dark forebodings which seemed to characterise this period, there was one bright spot for one section of the work-force. For some months, the trade unions had been negotiating with the company for increased holiday entitlement, which had remained unchanged for many years. This had culminated in the introduction of a three-week holiday for all monthly staff and for weekly staff on a graduated scale, depending on length of service. These changes would apply to holidays taken in 1958.

However, the earlier gloomy forecasts were seen to be only the forerunner of what was to follow. At the end of 1957 it was reported that expenditure had exceeded income by an average rate of £22,000 per week and the Directors took urgent steps to right the situation. These included a further loan from both parent companies of £100,000 in January 1958, and a reduction of stocks and capital expenditure in the coming year. At the same time, the Technical Director, L.G. Fairhurst, sounded a warning that unless there was an early increase in the number of new projects there would be surplus manpower capacity in the Technical Departments by 1959.

The Directors also had second thoughts on a decision they had made in October 1957 to develop a propeller in the 4,000 bhp field, when they learned from LGF that the cost of development to Certificate of Airworthiness standard would be in the region of £500,000, and that De Havilland would be firmly established with their own product by that time. The opinion was expressed that Rotol should concentrate on the 1,000 bhp range, where there appeared more likelihood of sales.

On February 19th 1958, a meeting of representatives of the parent companies and Rotol was held in the London office of the Bristol Aeroplane Company, in order to discuss future policy. (It is possible that this was the meeting referred to by Gill in October 1957, but if so it had taken a long time to arrange.) Although it was not altogether unusual for meetings to take place away from the Main Works, there may have been a special significance in this particular meeting. No written record has been made available to the author but at the next board meeting held in April 1958, the recommendations put forward by that special meeting were in full swing.

These recommendations became known as 'the retraction policy', a term which meant cutting back in a number of ways, but chiefly in personnel. By the end of the year a total of around 1,100 had faced up to the redundancy which the company imposed, practically all of these being 'direct' labour – the men and

Ron ('Steady') Barker of the British Messier Service Department at the wheel of his three-ton 1907 fire engine.

Stuart Docking testing one of the first of the 6-way fuel flow proportioners for the Handley Page Victor bomber.

The Managing Director of British Messier. L.S. Armandias (centre) shakes hands with the British Ambassador, Sir Gladwyn Jebb, at the Paris Air Show in 1955. Vernon Christie is on the right.

British Messier employees visit the Bristol Aeroplane Company at Filton on 10th October 1955 and seem well-pleased with their handiwork on the Britannia nosewheel undercarriage

British Messier design team in 1959. *Left to right*: Roy Fairclough, Alan Cameron-Johnson, Geoff Collinson, Bob Preece, Reg Brown.

Alan Cameron-Johnson and Charlie Davey share a two-man drawing board!

Cyril Luby
Managing Director

Bert Jelley
Works Engineer

Bill Nottingham
Production Engineer

Ernie Smith
Chief Inspector

'Mac' Cracknell
Despatch Superintendent

Bill Savory
Turbo-Charger Manager

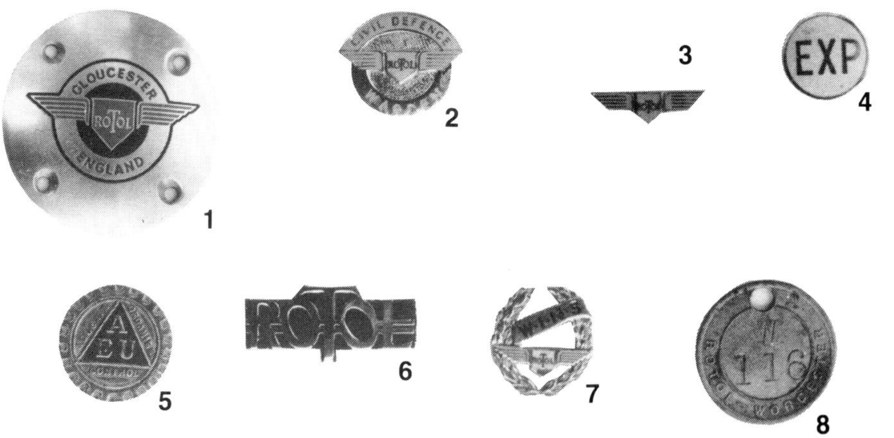

Examples of some Rotol Badges. 1 Spinner noseplate, 2 Civil Defence Warden, 3 Rotol 'wings', 4 Experimental badge, 5 AEU lapel badge, 6 Security badge, 7 WINS (Women in National Service) lapel badge, 8 Worcester Factory badge.

An example of a Rotol gearbox in the early 1960s. This one was for the Vickers Vanguard.

One of the original slide rules used in the Stress Department.

Cricket match at Bristol in 1951. Rotol Executives versus BAC. Seated in centre is L.G. Fairhurst and on his left is Dr A.E. Russell, BAC Chief Designer. Several other well-known Rotol personalities are also present.

Rotol Rifle Club *c.* 1955. *Left to right*, seated: Edden, Smith, Byworth, Newport, Chilman, Hart, unknown, Standing: Williams, White, Turner, Savory, Turner, Atkins, unknown. Grant, Turner senior, Brooks, unknown.

Probably Rotol's greatest post-war triumph. The Rolls-Royce Dart propeller on a Capital Airways Viscount.

The Executives of Rotol Incorporated in 1958. Jack Staples, President, Sir John Evetts, Chairman and Vernon Crudge, Director with a model of the Fairchild F27. (Licence built from Fokker.)

In 1957 Mrs Annie Arch receives a bouquet from Charlie Homer on the occasion of her 78th birthday. She joined Rotol in 1943 at the age of 64 and was still working in the Contract Loan Store at the time this picture was taken. The Personnel Manager, E.F. Yeomans, looks on at the right.

Farewell to Terry Elmes in 1971. *Left to right*: John Hunt, Jack Beach, Dick Watts, Eric Stockdale, Dick Smith, John Chilman, Tony Evans, Ian Golesworthy, Terry Elmes, Pete Spence, Arthur Lane.

The Annual General Meeting of Rotol Limited was held on July 24th 1958 together with a Directors' luncheon to celebrate the firms's 21st Birthday. Some of the distinguished guests present were: *Left to right*, back row: W.A. Robotham (Rolls-Royce), J.D. Pearson (Rolls-Royce), D.R. McDonald (Rolls-Royce), W. Masterton (BAC), Dr Llewllyn-Smith (Rolls-Royce), Captain K.J.G. Bartlett (BAC), Air Commodore F.R. Banks (BAC), A.F. Kelly (Rolls-Royce), Colonel C.W. King (Rotol), R.R. Raff (Rotol). Front row: W.T. Gill (Rolls-Royce), Rear-Admiral Sir Matthew Slattery (BAC), Air Chief Marshal Sir Alec Coryton (BAC), Lieutenant-General Sir John Evetts (Rotol), Sir Reginald Verdon-Smith (BAC), Captain C.W. Uwins (BAC), C.J. Luby (Rotol), Dr A.E. Russell (BAC).

Sir George Dowty addressing the Rotol and British Messier apprentices annual prize-giving in 1959

An aerial view of Rotol at the time of the Dowty take-over. The firm of Wynstruments is in the foreground and the 'White Block' now carries its third storey. The six large factory buildings on the near (south) side of the road were RAF hangars during the war.

women who manufactured the products or 'those who turned the handles' to use a favourite phrase in the personnel department at that time. (The ratio between 'direct' and 'indirect' labour – the latter were mainly the clerical and technical staff – was always a matter for close scrutiny by the directors and any sign of imbalance was cause for concern.)

This cut-back in the work-force was doubly unfortunate. The company had been worried for some time at the steady loss of machinists and had made great efforts in the previous months to build up the numbers of skilled workers – who were always at a premium in the district. Unfortunately, accurate figures of the Rotol and British Messier work-forces throughout the years are not known, but occasionally they are quoted in the Board Meeting Minutes. In August 1956, the direct labour force totalled 1,707 of whom 1,269 were employed in Main Production, the remainder in the Toolroom, Experimental and Research Departments. By the end of 1958, after the redundancies had taken place, the total number of employees was around 4,350. This figure is believed to include all Rotol and British Messier employees, both staff and works.

Staff reorganisation following the redundancies put 'Jock' Maclean into the Works Manager's seat with Reg Barrell as Chief Inspector. Ernie Smith became Deputy Sales Manager, with a specific brief to look into the over-provisioning of spares for civil aircraft equipment, which was giving cause for concern as mentioned above.

For many years it had been standard practice at Rotol to hold Board of Directors meetings at six-weekly intervals, at which the first item of business was the election of a chairman, from either Rolls-Royce or the Bristol Aeroplane Company. At the one hundred and fifty-fifth meeting, held at the offices of the Bristol Aeroplane Company, 6 Arlington Street, London, on June 25th 1958, this custom was broken when it was unanimously agreed to appoint Sir John Evetts as Chairman of the Board of Directors and C.J. Luby to Managing Director. These appointments were to become effective as from July 1st 1958. The decision to appoint a full-time chairman was undoubtedly a welcome one, although some might argue that it could have been taken at a much earlier stage. With hindsight it was a curious time to make the change. The end of the road for Rotol and British Messier lay just a few short months away, and the secret talks with the Dowty Board must have been imminent, if indeed they had not already begun.

These changes of the Board of Directors coincided with the Twenty-First anniversary of the company's formation, and Rotol were poised to celebrate this with plans for a fairly modest function, but these had to cancelled because of the financial situation. Fortunately, the laying down of the new bowling green, which had been planned for some time, went ahead, and became a welcome addition to the facilities offered by the Sports and Social Club. The green was a gift from the management to mark the company anniversary, and was officially opened by Sir John Evetts and C.J. Luby on the following Whit Monday.

One other event connected with the 21st Anniversary Year took place on Saturday, July 13th 1958. A Grand Carnival was organised under the able leadership of Vic Baguley, the Secretary-Manager of the Sports and Social Society, and a team composed of Jack Shea, Jack Shaw, Steve Cerrone, Ian Golesworthy, Tom Lacey, John Bartleman, Jack Langston, Bill Sykes, Freddie

Day, Jack Tallett and John Luscott, with Bert Jelley as chairman. A marquee was erected on the Sports Field and a display of Rotol products exhibited within, ranging from early war-time airscrews to the new turbo-chargers and fuel flow proportioners. Outside was a roped-off arena, scheduled to feature non-stop entertainment during the day and in addition there were side-shows, roundabouts and a fun fair to complete the day's events. The Rotol Carnival Queen, Diane Carr, together with her attendants, Kathleen Joyner and Christine Tiller were present in addition to a number of civic dignitaries and Rotol executives and their wives. Due to inclement weather, the outdoor programme – the firework display, dancing on the grass and beating the Retreat by the Band of the Gloucestershire Regiment – had to be cancelled but despite the rain, everyone seemed to have enjoyed themselves and it was a welcome respite from the gloomy news of the time.

A possiblity of new propeller business came on the scene during 1958, when the Armstrong-Siddeley Motor Company put forward proposals for a new engine in the 1,000 hp class, known as the P.182. It was agreed to press ahead with a brand new self-contained design of propeller, (there was no separate oil supply to the propeller) which would be suitable for both the P.182 and also an engine that Lycoming, an American company, was hoping to establish in Europe. The argument was advanced that Rotol could not afford to ignore this market, even though the estimated cost of the venture was put at £150,000, but due to patent and other difficulties, the project was abandoned at an early stage.

Ways and means of further economies were considered at the October meeting of the Board, including the transfer of British Messier Technical and Experimental Departments to the North Works, and the possible addition of a second floor to the Technical Block. These transfers would enable the South Works to be closed completely, a move which had been forecast some months previously. The situation was becoming more difficult with each day that passed and the Rotol senior executives were deeply concerned. Although the manufacturing shops had sufficient work for the immediate future, the same could not be said for the Design and Development departments. It is is reasonable to assume that during this period, the joint boards of Rolls-Royce and BAC came to the conclusion, possibly with Lord Hives as the key figure behind the move, that the only sensible course open to them was to sell off the company. Behind the scenes, secret discussions began in an effort to find a buyer. The choice was perhaps, in retrospect, an obvious one.

George Dowty had started in the early 1930's as a manufacturer of aircraft landing gear and over the years had diversified into many different fields of engineering. By the end of the fifties, Sir George Dowty (he was knighted in 1956) had built up a thriving industrial empire, and the acquisition of Rotol and British Messier must have seemed an attractive proposition to complement his original company, Dowty Equipment, who were still manufacturing aircraft landing gear. After some hard bargaining, terms were agreed for the sale of Rotol and British Messier to the Dowty Group of Companies, whose head office was at Arle Court, only a couple of miles away. The main stumbling block had been the large amount of stock and work-in-hand at Rotol, but once a price had been negotiated the deal went ahead. The author has it on good authority that the

Dowty Group acquired the Rotol and British Messier shares for a sum of £2,235,000, giving Rolls and Bristol over one million pounds apiece. In the light of subsequent events it would appear that this was something of a bargain, as considerable profit was to be made in the 1960's from the Dart propeller and gearbox equipment alone.

One of the men most closely involved with the talks which had taken place behind the scenes was Mr R.F. Hunt, (later Sir Robert) and he was able to provide the author with a few more details concerning the actual discussions leading up to the take-over. He gave it as his opinion, that Rotol's parent companies, Rolls-Royce and BAC were finding it increasingly difficult to maintain an amicable working partnership, due in part to the fact that they were still in competition with each other, particularly in the field of gas turbine engines. Also, they were by no means convinced that a secure future lay ahead for aircraft propellers, despite the promise and sales of the Rolls-Royce Dart turbine engine. The undercarriage and hydraulic business, taken over from British Messier, was not their main product and neither Rolls nor Bristol were particularly interested in promoting it.

These comments by Sir Robert are only one of a number that have been made over the years. There have been a great many rumours and much interesting speculation on the reasons for the Dowty take-over of Rotol and British Messier. Older employees have given several differing versions of the 'truth' as they saw it which lay behind the action, some of them positively scandalous, but if there is any more than the factual account which follows, it has been a well-guarded secret for over thirty years. From the documents which have been made available, it appears to have been a simple agreement of the two original parent companies making a decision to sell-off their asset in order to allow them to concentrate on the problems of running their own businesses. The aviation industry, like most others, has always had its ups and downs, and both Rolls-Royce and BAC were experiencing their share during the time of these events.

It was at the hundred and sixtieth meeting of the Rotol Board of Directors, held on Thursday 18th December 1958, that the announcement was made, at which the following were present: Sir John Evetts, L.G. Fairhurst, C.J. Luby and Col C.W. King, all Rotol Resident Directors and the Nominated Directors from the Parent Companies: W.T. Gill from Rolls-Royce and Air Commodore Banks and W. Masterton (who had been invited to attend in the absence of Sir Reginald Verdon-Smith) representing the Bristol Aeroplane Company. The company secretary, R.J. Raff was also present, and there were apologies for absence from Sir Reginald Verdon-Smith and Whitney Straight.

The minutes of the meeting commenced with a prepared statement by Mr Gill, in which he informed the Board of conversations that had taken place between Sir George Dowty and Sir Reginald and himself, representing the shareholders in Rotol Ltd. As a result of these conversations, Rolls-Royce and Bristol Aeroplane Company had agreed to sell their shareholding in Rotol to Dowty Group Ltd. The statement went on to record that it had been appreciated for some time that new work in fields outside those in which Rolls and Bristol were directly interested, was needed by Rotol for the future. It was hoped that this step would be in the best interests of all those concerned at Rotol and that a joint announcement would be made to the Press the following day.

As would be expected, the minutes only record the bare bones of the meeting. No mention is made of any discussion, indeed, according to L.G. Fairhurst, the Rotol Resident Directors were all taken aback by the finality of the statement and very little was said. It may be that Sir John Evetts knew of the talks going on behind the scenes, but if so he kept tight-lipped about it. The meeting adjourned for lunch, where they were joined by Sir George Dowty and Mr R.F. Hunt, Dowty Group Deputy Chairman. Lunch was a somewhat subdued affair, with not much light-hearted discussion and at the conclusion, Sir George expressed the hope to those present that a satisfactory future lay ahead for the Rotol organisation as a member of the Dowty Group.

The following day, Sir John Evetts made an announcement concerning the sudden turn of events to a large gathering of his most senior executives and managers in the canteen, and from this meeting the news travelled swiftly around the offices and factory. In most places it was greeted with incredulity, but there were plenty of minor prophets who had 'seen this coming for some time' – or so they claimed.

The next few months was a busy time for the financial departments in particular, as the stock position had to be assessed quite accurately, budgeted profits and the bank overdraft and other like matters had to be considered. An Extraordinary General Meeting had to be called for the purpose of arranging the sale of shares and this was done on 26th March 1959. (The Rotol Resident Directors would not be present at this meeting, it being confined to the Rolls-Royce and Bristol directors.)

Rotol and its subsidiary British Messier, and Dowty Equipment, which had continued to operate as separate companies since the announcement of the take-over at the end of 1958, merged to become known as Dowty Rotol Ltd as from April 1st 1960. In a reversal of the more usual way of a take-over, the smaller company swallowed the larger. Dowty Equipment at Arle Court moved to the Staverton site and an amalgamation of departments took place, with inevitably some personnel becoming redundant. There was also a certain amount of down-grading at Departmental Manager level, where two people held similar positions, and this sometimes resulted in bitter words being spoken – not all of them in private.

A story which went round the factory at that time, prompted perhaps by the choice of date for the announcement, said that an alternative name for the company had been O'Liery Ltd – based on the 1937 formula of RO from Rolls and TOL from Bristol. In this revised version it was suggested that one took the ROT from Rotol, the MESS from Messier and the DOWT from Dowty to end up with O'Liery! Whatever else may be said of Rotol, there was always time for a spot of humour.

No wake was held for the airscrew company with its roots in Llanthony Road, Gloucester. Most of the workforce from Arle Court and Staverton simply got on with the job in hand, ignoring the changes which were imposed and in time, whether you were a 'Rotol' man or a 'Dowty' man became largely irrelevant. In some ways it was a re-run of the 1940's when if you were from Bristol you might not see eye-to-eye with someone from Derby and vice versa. The phrase, 'This is how we used to do it at . . .' was heard once more in the land, although the number who could remember the first time these words had been spoken grew steadily smaller with the passing years.

Today, Dowty Aerospace Gloucester as it is now known, continues to be at the forefront of aviation technology. Its gaze is firmly fixed on the beckoning horizons of the Twenty First century and the company will no doubt be well represented on all the new aircraft that lie in the future, just as it has always been in the past. Low noise, high-efficiency swept blades made from modern carbon fibre and glass-reinforced plastic are beginning to elbow aside the duralumin propeller blade and the humble graphite pencil, which once produced the thousands of drawings still to be found in the Negative Stores, has given way to the mysteries of computer-aided design. Young men and women who have never heard a siren, except the one which tells them the working day is over, now continue the story that began in an old carpet factory by the side of Gloucester Docks.

British Messier

The history of the firm of British Messier deserves a book devoted solely to its exploits and perhaps someone with a far greater knowledge of the company than the present author can be prevailed upon to write the whole story one day, but for the present we will have to content ourselves with this single chapter. It is regrettable that the Board of Director's book concerning the period from 1946 to 1956 is missing as it would no doubt yield much of interest concerning the various changes that were made during this time, enabling a much more complete account to be given of 'behind the scenes' at both Rotol and British Messier.

Although the part played by British Messier in the Rotol story is most important the exact point at which to begin presents a problem, as we are faced with the familiar difficulty of insufficient documentary evidence. Fortunately, several of the people who were involved in its origins have come to my assistance and have led me through a complex maze of events concerning the creation of British Messier.

We need to go back once more to the pre-war years to meet the man who has crowded more excitement into his lifetime than most. This man is Louis Sylvio Armandias and to catalogue his exploits would take a book in itself – but with luck he will accomplish this himself during the next few months. A very brief account of his activities is necessary to ensure an adequate understanding of the role he played at that time. He arrived in England in 1936, with a licence from Messier of France to produce their equipment in the UK and set about the task of finding a customer and a factory. He was successful in persuading Handley Page of the merits of the Messier system, which included not only the landing gear but also an innovative hydraulic system, which was incorporated in Handley Page's tender for Specification P13/36, the aircraft that eventually became the Halifax, the second of the RAF four-engined bombers of the Second World War.

Armandias had not at this time any design or production facilities but following discussion with Baron D'Erlanger and Alan Goode of Petters, he formed his own company with offices at Grosvenor Gardens, London. In order to obtain the required manufacturing capacity he joined forces with Owen, of Rubery Owen and became Rubery Owen Messier, with a factory in Warrington. Handley Page had been awarded a contract for two prototypes which included the Messier undercarriage and hydraulic system in the proposals. At this stage, April 1937, the design was for a twin-engined aircraft but because the RR Vulture engine was causing a great deal of concern, Handley Page set about redesigning the bomber to take four Merlins. A similar story was being enacted by A.V. Roe with their twin-engined Manchester – soon to become the Lancaster. It took some time for the Warrington factory to get onto a production footing and in the meanwhile, to avoid any hold-ups, the Air Ministry instructed the Dowty company to produce a levered suspension undercarriage. This was fitted to the Halifax B Mk V and was first flown on a Mk I, all other marks being fitted with the Messier design. Armandias was first the Technical Director and then Managing Director from 1939 to 1942, and his associates were messrs Conway, Orloff and Collinson, all of whom were to continue with British Messier in the years ahead. There is a strong body of opinion that the salesmanship and enginering competence of these three individuals played no small part in the creation and survival of British Messier.

In 1942, Armandias parted company with Rubery Owen and joined Army Intelligence. Little is known of this period of his life, indeed he has refused to supply the author with any details, possibly because he considers that the Official Secrets Act still covers his activities. He admits that he used two names and wore two uniforms which suggests that he may have been involved with clandestine operations in France, possibly with SOE – the Special Operations Executive responsible for planning raids and infiltrating agents into Nazi-Occupied Europe – although this is purely speculative. It is known that he was at Beaulieu during the war, which was then a centre for SOE training.

In the summer of 1941, the Rotol Controller, R.H. Coverley, probably acting in his capacity as advisor to Lord Beaverbrook, was engaged in discussions regarding the production on a sub-contract basis of a Messier Hydraulic Pump by Rotol. This led to his being offered the post of Joint Managing Director with Rubery Owen Messier but after discussions between Coverley and the Rotol Board of Directors the idea was abandoned – although what inducement was made to stop Coverley from pursuing the position is not known. Later in the year, although Rotol had made considerable progress with development and testing of the hydraulic pump, all connection with Rubery Owen Messier was dropped.

Armandias joined the newly-formed mixed helicopter squadron in 1943 at Feltham, operating four Sikorsky R4 helicopters that came to England in that year. By the following year he held Helicopter Pilot Licence No 5 in Britain and No 3 in France, after which he undertook a survey for the Bristol Aeroplane Company which resulted in BAC (now part of the Westland Group) setting up its own helicopter division.

His career took another turn in 1947 when he became the co-founder and managing director of British Messier Ltd with the help of Baron d'Erlanger and also Sir Reginald Verdon-Smith. Bristol Aeroplane Company held 50% of the shares of this new company, the remainder being held by Arbo Investments.

Hugh Conway, an associate of Armandias for many years, was appointed Technical Director. Those who were closest to him considered him to be one of the best technical directors in the business, tough but scrupulously fair and always approachable. He demanded a great deal from his people but gave back as much in support. Conway's influence, together with support from a strong technical design team led by George Orloff as Chief Designer and Geoff Collinson, Chief Technician, was to provide a high standard of engineering within the British Messier organisation.

British Messier began operating from some wood and asbestos huts at the

Cheltenham end of the Rotol Main Works in 1947, on the site currently occupied by the building universally referred to as 'Stalag'. The huts had been provided by Rotol and used for a variety of purposes prior to BM moving in, including a Post Office, a hairdresser and certain other amenities which were believed to be essential for morale purposes during the war, when little spare time was available for the work-force to go into the nearby towns. When BM moved in they turned one of the huts into the Design Office and used others for assembly and experimental testing, including the first brake test machine using a Ford six-cylinder engine.

The BM Design Office in 1947 comprised a surprisingly youthful team, Conway was only 35 and most of the draughtsmen were still in their twenties. J.J. (John) Wilson was Chief Draughtsman initially, with a number of section leaders – among them Roy Fairclough, Alan Cameron-Johnson, Reg Brown and Frank Taylor – controlling a number of draughtsmen. There were also two draughtswomen, Rena Pickard, who married Bob Preece, the Chief Stressman, and Beryl Hall who was tragically killed in an air crash some time after leaving BM. Rena carried on well into the period of the Dowty take-over as one of the members of the hydraulic jack design section

Among the first employees was a young man who was to play a major part in most of the technical design work of the company. Alan Cameron-Johnson came to British Messier in 1947, after a spell with Sir Roy Fedden at Stoke Orchard, near Cheltenham, where he been part of a team engaged on a number of interesting projects. Among these was a revolutionary car, with an aluminium alloy body, powered by a three-cylinder sleeve valve rear engine, and another an aircraft engine, designed to be buried in the thickness of the wing. Two others he met at Stoke Orchard were Don Bishop and Vic Holberton, both ex-Rotol employees. Don had been with Hordern-Richmond for some time, as has already been related in a previous chapter but when the Rotol Directors decided not to continue with wood propeller blades after the war ended, Don was offered a job in the Marine Screw office at Main Works but decided against it, electing instead to try his hand with Fedden's new venture. When it folded up, Don went back to Hordern-Richmond, where he bacame involved with making parts for textile machinery from the 'improved' woods the company had developed. Vic Holberton was approached by Fedden while he was still in the Project Design Office at Main Works and went first of all to Benton House in The Park, Cheltenham, before moving to Stoke Orchard, where the prototype car was built and tested. In the event none of the projects achieved production status and Fedden wound up the business. Vic returned to the Rotol Project Office, until another approach from Fedden resulted in him joining a design trio which Sir Roy was infusing into a new Dowty subsidiary. This became Dowty Fuel Sytems and Vic went on to serve his remaining working years with that company.

Ray Base, an ex-RAF Flight Engineer on Stirlings before his demobilisation, joined the company in June 1947 – his clock number was 13 – and worked on a variety of landing gear, hydraulics and brake systems. He was section leader of first one, and then both landing gear/hydraulic sections in the DO and eventually took charge of the original BM Project Office. British Messier retained complete autonomy in design until the merger with Dowty Equipment in 1959, and among his last jobs was the design of the nose gear for the Avro 748, under Reg Brown's direction.

Cyril South, who came to British Messier in 1948, had some trouble finding the Design Office initially as the wooden signboard had rotted and fallen off the side of the shed. When he eventually located the office he joined the small number of draughtsmen and designers, who were probably only just into double figures at this period and early in 1949 he was asked to take on the job of test rig section leader. Until this time, undercarriage testing had been somewhat primitive but Conway was anxious to rectify this situation and from then on large test rigs were a feature of life at British Messier.

Charlie Davey (we last heard of him working in the Rotol Research Department during the war) returned to the site after a brief sojourn with the electrical firm of Smith's at Bishops Cleeve. He was involved with the undercarriage lock jacks for the revolutionary Short Sperrin, the first four-jet bomber, of which only one example was built. The very tall steel main landing gear for this aircraft was designed by BM and in common with current practice in those days was drawn on millimetric squared graph paper, and instead of 'hatching' a sectioned view, each part was lightly coloured, this being the practice at French Messier.

In those days Charlie Davey was noted for wearing rather 'loud' shirts and he recalled one occasion when Conway stopped at his drawing board, on his weekly tour of inspection. These were a feature of life in the DO when Conway and J.J. Wilson would look at everything that was going on and ask questions that the draughtsmen were expected to answer. Conway told Charlie that he was going to the United States in a few days time and would bring him back a suitable flamboyant shirt from the West Coast. Sure enough, Charlie was called in to Conway's office on his return and presented with the promised article. This little story tells us more about Conway and the esteem in which he was held by those who worked for him, than a whole chapter might convey.

The use of the huts by BM was merely a temporary arrangement until the site recently vacated by the Rotol Flight Test Department on Staverton Airport was converted to accommodate them, and they moved to the new premises in 1948. They found a vast difference in their new surroundings, with a good deal more space and light than they had previously enjoyed.

Among the earliest aircraft components designed and made at BM were the dive-brake jacks for the prototype Canberra bomber made by English Electric at Wharton. Special valves called flow-dividers were supplied with these jacks, ensuring that whatever the difference in air-loading on either wing, the dive-brakes always extended at the same rate and so did not upset the equilibrium of the aircraft. To prove that they really worked a special rig was made for the British Messier stand at the 1949 SBAC Farnborough Air Show, consisting of two jacks side-by-side, one with a large weight fixed to the top of the jack and the other having a small tray in which was placed a feather. Visitors were asked to operate the rig, and were amazed to find that both jacks extended at exactly the same rate.

At the rear of the converted Flight Test hangar, BM erected a large drop-test rig for undercarriages and also a machine for brake testing. The brake test department carried out a lot of development work using different materials and coatings for the discs and brake pads and the first set of disc brakes produced at BM were in fact fitted to an all wooden construction light aircraft produced by Elliott's of Newbury, who were more well-known as furniture makers. The Eon was a four seater low-wing monoplane with tricycle undercarriage, mainly used for business purposes and glider towing and never went into production.

In the late 1940's, under Conway's dynamic leadership, the company were busily engaged with most of the major British aircraft projects – and quite a number of minor ones too – producing landing gear and 4,000 psi hydraulic control systems requiring a completely new range of oil seals to withstand the higher pressure. All production manufacture was done by Rotol but British Messier had its own Experimental Shop, under Jock Maclean, a very popular Manager whose early career at Rotol has already been briefly described in a previous chapter.

One of the many characters at British Messier at this time was the pump test engineer Zdislav Ludwik Zeyfert – thankfully always known as Louis – who flew Staverton's Murray Chown Aviation eight-passenger Rapide at the start of the Channel Islands service. Louis, also a flying instructor for the Gloucester Flying Club, had had an exciting life having escaped from his native Poland at the time of the German invasion, served with the French Armee de l'air and when France was overrun, escaped again to Britain where he flew for the RAF on Halifaxes. When he left BM he flew Constellation airliners for the Dutch airline KLM and finally retired in Holland.

Changes were made in the management structure at BM in 1949 when Rotol acquired Bristol's 50% interest for £5,000 and subscribed for an increase in share capital by contributing another £20,000 in 1950. The remaining 50% was purchased from Arbo Investments Ltd for £50,000 in 1952 and the company thus became a wholly owned Rotol subsidiary. This 'take-over' by Rotol was presumably to regularise the position and put both companies on a more sound footing. The Board of Directors of British Messier in November 1952 was as follows; Lord Hives (chairman), Lt Gen Sir John Evetts (managing), L.S. Armandias (deputy managing) Hugh Conway (technical) and either Verdon-Smith or W. Masterton of the Bristol Aeroplane Company.

Guided Weapon work was introduced into the company in 1951, following a visit by Conway to the Royal Aeronautical Establishment at Westcott where work on solid propellant rocket motors was being carried out. They were having trouble with welding various sizes of tube from DTD124A (a relatively low-strength steel) and Conway offered to produce some samples at BM which he assured them would be satisfactory. This he did, in spite of having very little equipment or experience, and largely on the strength of this exercise he obtained a contract for development of rocket motors, which were variously powered by solid propellant, air or combustion gas as well as oil.

To cope with all this extra work, Conway promoted Cyril South from Section Leader (Undercarriage) to Project Engineer of the new product. The GW design section was under the leadership of Ian Lewin, and Rod Dale who died tragically young. At one time they had a complete airframe of a captured German Messerschmitt Me163 rocket fighter, in order to study its wire bound pressure vessels. Much of their work was on the Avro Blue Steel weapon, which was cancelled in 1956. P.P. (Pat) Robinson was made Chief Engineer Rocket Motors and Frank Wilkinson, from Metropolitan-Vickers in Manchester joined BM as Welding Research Engineer. J.G. (Jim) Martin transferred from Service Engineer on the aircraft side to assist with welding all the early motors until they were eventually taken over by the Bristol Aeroplane Company and subsequently Bristol Aerojets. This move occurred following some altercation between Hives and Conway, when Hives is reputed to have protested most strongly at British Messier moving into this field. The end result was the closure of the Guided Weapons Department, with Conway leaving for Short's of Belfast, where guided weapons were both tolerated and prominent, and a number of others from British Messier went with him.

Gordon Gardner was part of the Guided Weapons Team in the DO and at the time of this next tale, 'Mick' Thurston had just taken over the post of Chief Draughtsman. It had long been the custom, both at Rotol and British Messier, for anyone who was getting married to be fitted with a ball and chain, worn on the last working day before the wedding. Thurston was conducting an interview with a junior applicant when Stan Stevens, who had just received this emblem of the last day of bachelorhood, staggered into his office (slightly under the influence of a recent excursion to the club house) and pleaded, 'Sir, I've finished that job you gave me, can I have the chain off now?'

The Bristol Britannia undercarriage was one of the most complex units ever produced by British Messier up to that time. It certainly had more than its fair share of difficulties in the early days because of a fatigue problem which took a great deal of solving. The main gear was designed jointly by R.S. (Dickie) Ransom and Alan Cameron-Johnson. Dickie started the initial design and laid out the basic priciples which were something of a novelty at that time. The bogie, with its four wheels, had to retract backwards into the engine nacelle, at the same time rotating to fold alongside the main leg, before a sequence valve signalled the leg itself to retract. The nose gear, with a sophisticated power steering system and two wheels coupled on a rotating axle to suppress shimmy, was the responsibility of Reg Brown and Ray Base.

A jig borer named Wally Bullock was working on the Britannia main fitting and he took a great deal of trouble to ensure that each of the holes he was asked to produce was machined to the best of his ability. One hole in particular caused him some difficulty and when he took the opportunity to go to Filton to view the aircraft, on a visit arranged for the workers who were engaged on the project, he took note that 'his' hole did not appear to be doing anything. When he enquired from the helpful guide who was conducting the tour, what was the purpose of the hole which had caused him so much trouble, he was disappointed to learn that its function was to carry a red marker flag, indicating that the undercarriage was locked in the 'Down' position. However, he cheered up when he was informed that it was one of the important pre-flight checks and would only be removed prior to the aircraft taking off.

Conway and his wife, Eve, lived in The Park (a residential area in Cheltenham) and Cameron-Johnson recalled the time he and a team from the Design Office spent one complete Sunday, thrashing out design problems and preparing an urgent submission for a customer. They were quite content with this arrangement, as the surroundings were a decided improvement on those at the office and Eve Conway – who was almost as commanding a figure as her husband – kept them liberally supplied with food and drinks. As already indicated, the relationship between Conway and his men was always a happy one.

Alan's progress from design draughtsman to Chief Undercarriage Designer at BM encompassed all the products with which the company were associated, ranging from simple valves in various hydraulic systems, to the undercarriages for the Britannia, Seamew, Gnat, and Lightning, as well as a number of helicopters. He and Conway were motoring to Folland Aircraft at Hamble, in connection with the Gnat undercarriage, with Conway driving his Sunbeam Talbot – a rather small car for his bulk. They were speeding along the road into Marlborough when Conway suddenly had an idea, but instead of stopping the car he held the steering wheel between his knees, and using it as a support, proceeded to jot his thoughts into a notebook!

The production contract for the Gnat undercarriage was produced by a subcontract factory at Stratford on Avon, Peak Engineering, as they had tooled up for this job and were well versed in the requirements. 'Jimmy' James recalled that a shift in company policy forced him, despite his protestations, to transfer the job to the New Mendip factory – which had neither the expertise or the relevant tooling to carry it out. Nevertheless the contract went to New Mendip and shortly after Jimmy received a telephone call from them asking for information as to the whereabouts of the previous contractor, which he gave them. The reader may guess the rest of this tale . . . New Mendip subcontracted the job to Peak Engineering who delivered the goods to New Mendip, who sent the finished articles back to British Messier. Everyone was happy, including Jimmy's boss who delighted in saying, 'I told you that New Mendip could do it!'.

Hugh Conway, already the Technical Director at British Messier, is thought to have cherished hopes of achieving the same appointment at Rotol, a post that was not in fact created until some years later. Possibly because of this, and some other behind-the-scenes controversy which took place between Hives and himself, already referred to, he severed his connection with the company and moved to Short's at Belfast in October 1953. There, he was Deputy Managing Director and Chief Engineer until 1964, then becoming Managing Director of Bristol Siddeley Engines from 1964 to 1968. Hugh Conway CBE was appointed to head the committee that introduced decimal currency into the United Kingdom in 1971 and died in November 1989 as this book was still being researched.

Probably the most complex undercarriage geometry with which British Messier were associated was the main gear designed for the English Electric P1 fighter which became the Lightning. It had to retract into a wing no thicker than about 10″ and because the wing itself had a 60° sweepback, the gear was required to swivel as well. The designer responsible for solving all these geometric limitations and the performance and structural aspects of the specification was Alan Cameron-Johnson, and Charlie Davey was appointed the BM Lightning Systems Engineer, both of whom worked closely with the aircraft design team at Warton.

Among the last of the aircraft to be produced by the Bristol Aeroplane Company before its merger with British Aerospace was the all stainless steel research aircraft, the B.188. Ray Base of the BM design office was responsible for designing the main landing gear, and Ivor Franklin, in the recently opened London office designed the nose gear. The thin wing of the B.188 meant that the complete wheel and axle rotated 90° during retraction, to enable the gear to stow in the wing with the wheels vertical in the side of the fuselage. Other research aircraft using landing gear designed by BM were the Short SB4 and SB5. The Short SB6 Seamew, which went into production for the Royal Navy in the late fifties, was somewhat unusual for the period in that its landing gear, designed by BM was not retractable.

BM maintained an active social side, with the usual tennis and cricket teams, and also a strong motor sport section. This centred around the BM Service Department where Jack French (who was last referred to in the Repair Department at Rotol during the war) and Ron 'Steady' Barker, together with John Marpole, and Ray Base all participated in many rallies and hill climbs during the 1950's. Barker, who was later to move into journalism and become associated with the world-famous Beaulieu Motor Museum, was the proud owner of a 1906 vintage motor car that had ben converted into a fire engine, and on one memorable occasion, brought the vehicle to work and drove a number of the lads to the Bat and Ball public house for a lunchtime drink.

Motor cars seem to have been very much a feature at BM according to several people who recalled those days. George Hinton, one of the Guided Weapons team, owned the original Wolseley Hornet show model, painted grey and in immaculate condition. It dated from around 1936 and is believed still to be in the possession of his family. Dickie Ransom was another who was noted for his car. He bought an old Austin 7 for almost nothing and made a sleek aluminium body for it, polished and unpainted in true aircraft fashion. When Alan Cameron-Johnson spoke to him in 1989 he told him that it had been sold to a dealer in Cheltenham, and later saw it advertised by the same dealer as 'an interesting trials special'.

A surprise announcement was made in March 1958 that Mr L.S. Armandias had terminated his association with Rotol and British Messier. When he left, Mrs Eileen Robjohn, (née Walker) who had become his secretary at the time of the merger with Rotol went with him, and continued to act in this capacity even though his subsequent career involved a number of various appointments in both finance and engineering, together with extensive travelling all over the globe.

Head figures at BM following the resignation of Armandias were: L.G. Fairhurst – Technical Director and Chief Engineer, E. Danvers – Deputy Chief Engineer and E. Morris – Chief Designer. Both 'Dan' and Ted Morris had been seconded from Rotol to oversee the new structure and from the outset both these Rotol men were received without rancour, partly because neither were regarded as company 'hatchet men'. E.G. (Geoff) Collinson was the Assistant Chief Engineer for a time but eventually decided to join Conway at Short's. Other heads were Alan Cameron-Johnson – Assistant Chief Designer, Roy Fairclough – Chief Hydraulic Engineer, E.N. ('Mick') Thurston – Chief Draughtsman, Reg Brown – Chief Undercarriage Engineer, Bob Preece – Chief Stressman. George Orloff moved to Rotol about the same time to run a new department, known as the Special Projects Section. As in the Rotol organisation, changes and promotions occurred from time to time, making it difficult to maintain a precise 'family tree' for either of the companies.

With the take-over by the Dowty Group in 1959, all the Technical Departments in the three companies, Dowty Equipment, Rotol and British Messier, were eventually brought together under one roof, in the completely refurbished building which was previously the Rotol Experimental Shop, where they continued as separate organisations, each maintaining their own products and loyalties. Whether this policy was conducive to ensuring efficient production remains the subject of some debate, but as it lies outside the scope of the present work it can be left to others to make their own judgements. It is an indication of the loyalty felt by those who worked for British Messier that a number of extremely successful reunions for past employees have been held, the first in 1983, another in 1988 and a third is planned as this book is being written.

Notes on Sources

Although countless references to Rotol may be found in the aviation magazines and periodicals over the years, and the company is frequently referred to in many books dealing with the Second World War, the author is not aware of any published work devoted to the history of Rotol or Rotol Airscrews. It was largely because of this that the present work was undertaken.

An important milestone for Rotol (and a fortunate one for anyone researching the history of the company) came in 1940, when the first copy of a brand new magazine devoted to the chronicles of Rotol, its employees and the dispersal sites came into being. It had a shaky start, as do so many ventures of this sort, but it settled down quickly and went on to appear right through the war years. *The Airscrew* at the outset was a semi-glossy publication priced at 8d., published by the Welfare and Social Society and contained a mixture of technical articles, sporting news, short stories, cartoons and news from the factories. At first there were a few pages of advertising but these disappeared later in the war when it became necessary to conserve paper and the page size was reduced. Because of its smaller size the price was dropped to 6d., but surprisingly the quality of the contents improved beyond all recognition.

It was apparently well received in some circles, as an early issue speaks of copies being retained in the archives of the Imperial War Museum as a typical example of a wartime house-magazine. Copies were also deposited with the local libraries, but only those in Gloucester seem to have survived and this is not quite a complete set, although they are all in pristine condition.

The first few issues were compiled by a committee, with first Ken Thatcher and then Guy Marson in the editorial chair, although the sub-editor was Tom Cruwys – who was mainly responsible for the contents. The Rotol Controller, R.H. Coverley (who seems to have had a penchant for titles!) was the 'Editor-in-Chief', presumably in order to have the power to veto anything which might not be to his, or the managements, liking.

Tom Cruwys was a superb artist and examples of his pen and ink sketches abound in the pages of *The Airscrew*. Prior to joining Rotol in 1939, he had enjoyed a chequered career, starting with a stage debut as a singer in the early years of the century, when he appeared on the same bill as Fred Karno and Charlie Chaplin. Forsaking the lure of the footlights after his voice broke, he became an art student but at the outbreak of the First World War on August 4th 1914, he joined the Honourable Artillery Company. During a fierce action on June 15th 1915, Tom was one of the few survivors when his battalion of over 1,000 officers and men were reduced to a mere 190. He next took a commission in the Worcestershire Regiment, was wounded three times and eventually 'demobbed' in 1919. Over the next few years he worked as an artist for interior decorators and then did free lance art work, before coming to Rotol.

Photographs in *The Airscrew* were of excellent quality, the majority being taken by Don Williams who joined the company in the summer of 1939. He had been with Burrow's the printers in Cheltenham and was persuaded to come to Rotol by another Burrow's employee, Mrs Durham. She told Don that when she applied for a job with Rotol she was so astounded at the difference between her old salary and the present offer that she had to take a firm grip on the sides of her chair. With this assurance he followed her soon after and started the Photographic Department which at that time came under B.W. Bleaney, the Publicity Manager, with Mrs Durham as his assistant for many years.

Tom Cruwys was largely responsible for producing the magazine and ensuring its distribution from the editorial offices at firstly 'The Lawn' in Gloucester, and then 'Glensanda' in Cheltenham. By the time the war ended, over 20,000 copies of *The Airscrew* had been sent, free of charge, to the men and women in the Forces, by the Rotol Welfare Office. They had also despatched over a million cigarettes, in the days when smoking was not only popular, but actively encouraged!

The magazine had, of course, to reflect the official viewpoint, either of the company or the government, and for this reason it tends to give a one-sided view of events. This is only what one would expect, but it is useful to bear in mind when considering the opinions expressed. As an example, *The Airscrew* refers to the Friday lunch-time entertainments held on the stage in the works canteen, and paints a pleasant picture of the audience sitting enthralled, applauding each item with gusto. D.M. Hill also mentions these concerts in her book (see below) as 'pretty mediocre but nevertheless much looked forward to as a break in the routine'. As in most things, the concerts held in the canteen, whether put on by ENSA or the Rotol personnel, had their good and bad sides, but it is probably true to say that they were less polished than a West End production. It would be unfair to put too much emphasis on this point, as ENSA (Entertainments National Service Association, but more often referred to as 'Every Night Something Awful') was generally looked on as a welcome relief in time of war.

Arnold Jackson, the Works Manager at the time, gave what is perhaps an accurate summing up of *The Airscrew* in November 1942:

> I am beginning to like this book now. Prose, poetry, pictures and people portrayed by persons that I can talk to. Persons that I work and play with. I feel on intimate terms with it and one can only be intimate with friendly people and friendly sentiments. A pleasant book, created by pleasant people, for the hour of slippers, dog and fire. That surprising hour when, as you read, you suddenly find the author's name – you never realised that he or she could do it. Your regard for these people automatically becomes something solid, because for some brief moment you were pleased and interested. That is why I am beginning to like *The Airscrew*, because it is increasing my regard for other people, who can give their precious leisure for our enjoyment.

Jackson wrote this before the magazine had been reduced in size by wartime regulations. The smaller size restricted the contents considerably, with little space for any of the record or film reviews, short stories, many written by Stanley

A pen and ink sketch by Tom Cruwys, Publicity Department.

Jackson the sub-editor, and – thankfully – the rather tedious diatribes about the enemy and similar pompous articles, so typical of the period. (Wartime propaganda seems amazingly puerile in hindsight but presumably helped to keep up the spirits of the readers.) In their place were many interesting accounts of visits to various departments, both local and further afield, in which the workers are named and often portrayed, usually by the camera of Don Williams. Rotol Main Works figured prominently but the dispersal factories were also encouraged to make sure they were well represented, submitting short accounts of activities at places like Frome, Worcester and Ossett.

Over fifty monthly issues of *The Airscrew* were produced and finished with issue No 53 at the end of 1945, although no mention was made in this last magazine that publication had ceased. Its pages contain a wealth of material for future historians, and it is hoped that a complete set will eventually be found, to be held in the County Record Office. They are a rich source of events that took place during the war years, and I have drawn extensively on their contents in order to compile this history.

Ten years after the final issue of *The Airscrew* another high-quality house magazine began its short career. This was the *Rotol and British Messier Quarterly Review*, the first issue of which was dated January 1955. In just under five years a total of twenty issues were produced, once again under the editorship of Tom Cruwys, with photographs by Don Williams. The format was somewhat different, as might be expected, but the magazine had many excellent articles and provided the employees with a record of life at Rotol and British Messier in the 1950's. Publication ceased after the take-over by the Dowty Group, which already had an excellent house magazine. A number of employees have loaned individual copies of both publications for the two periods, for which I have been most grateful.

Two other books purport to deal with Rotol life during the war, the first written by a lady who worked for two years in the machine shop. During her time with Rotol, Diana Murray Hill observed the lives of others and after she left was persuaded to put down her recollections, thus enabling her to publish a book in December 1944, entitled *Ladies may now leave their machines*. The title came from an announcement made over the tannoy system which gave the women a three minute start on the men in order to allow them an unhindered passage to either the main gate and the waiting buses or the canteen, a practice that still continues at Dowty Rotol. The book, with some 130 pages, was published by the Pilot Press Ltd 45 Great Russell Street, London price 7s. 6d., (37p) and is long out of print. It received some good notices from many of the reviewers, most of whom praised its straightforward approach for offering no concessions to the more usual story of factory life during wartime.

Diana Murray Hill was one of the thousands of women from all over the country who were required to work in support of the 'War Effort'. If the book accurately portrays her career, she came to Rotol in 1942 after some weeks at a Government Training Centre in Gloucester, and commenced work with Rotol ('QAD' in her book) staying with the company until she obtained her release on medical grounds in 1944. We are not told anything of her earlier life but from people who knew her I learned that she was sister-in-law to the British film actress, Phyllis Calvert, from whom I obtained her present address. When I contacted her she kindly filled in a few details concerning the book which intrigued me and gave permission to make reference to its contents in certain passages of the present work.

It must be said that Diana Murray Hill does not appear to have enjoyed the time she spent at Rotol, nor does she have any high regard for the majority of men and women she met – indeed in recent correspondence with the author she admits that her lasting memories are of the workers seemingly eternal quest for money and the ghastly pale-yellow lighting in the workshops. Few of the characters she writes about have any admirable qualities, and much of their time is spent in fruitless romances or endless bickering. Her two years with the company was in direct contrast to what most people remember of the days they spent at Rotol, as almost without exception most of the people I have spoken to assured me that their memories were happy ones, whether of wartime or of more recent years. One suspects that her background was light-years away from factory life, as it probably was for the majority of women in wartime, and she admits that she had very little aptitude for operating the machines, breaking taps and drills by the dozen as well as being a constant source of irritation and worry to her setters and supervisors.

There is the usual disclaimer at the front of her book which states that 'the characters are imaginary, although based on first-hand experience', but several people have told me that they recognised the real people lurking behind some of the names. Certainly, anyone who was at Rotol during the days of which Miss Hill writes would enjoy the book, with the proviso that the reader must make due allowance for the fact that she could hardly wait for the day when she would be released from its clutches.

She received a brief mention in the pages of *The Airscrew*, when she took the part of Mrs Bramson in the Emlyn Williams thriller, *Night Must Fall*, at the Opera House in Cheltenham, now the Everyman Theatre. This was a production by the Rotol Amateur Dramatic Society, and according to the reviewer, 'Diana Murray Hill displayed all the irritabilities of an eccentric old lady, admirably suited to the character'.

The famous writer and broadcaster, J.B. Priestley, spent some weeks at Rotol and was closeted with Miss Florence Jackson, the Controller's personal secretary, for part of the time. He was researching background material for a novel of the British people during wartime, set in a Midlands aircraft factory and entitled *Daylight on Saturday*. (The title alludes to the prevailing conditions, when the only natural light the factory workers could enjoy was at the weekend.) It was published in 1943 by William Heinemann and anyone familiar with Priestley's work will almost certainly enjoy this book as a fine example of his forthright style of writing. How the characters in the book compare with those he encountered at Rotol is difficult to judge but Priestley obviously made use of the experiences gained at Rotol and the other factories he visited.

Appendices

Appendix 1: Rotol numbering system

The Rotol drawing number system used a simple RA (Rotol Airscrews) prefix commencing with RA.100 – a Bolt, (proving the old adage that 'great oaks from little acorns grow'). This drawing, dated 15th June 1937, is part of the internal cylinder variable pitch propeller, and a microfilm copy can still be found in the present Drawing Stores. Not every original drawing has survived but a great number have fortunately been microfilmed. The oldest original drawing that can be located, an ink tracing on linen, is RA.162 Retaining Ring.

When the company started to produce accessory gearboxes they adopted a G prefix for these drawings; G.013 Oil Sump Group (drawn 8th July 1940) is still in existence, and it is assumed that the series began with G.001. Draughtsmen merely took the next available number from the register, in either the RA or G series, regardless of whether the part was a nut or a large casting. The highest numbers used in this system are RA.70553 Hub Assembly (for a two-blade ram-air turbine) and G.33125 Plug, both drawn on January 31st 1961, some two years after the Dowty take-over.

Reference to a series of drawings originating from the Bristol Aeroplane Company with FB numbers occur in early company documents. The reason for the choice of FB as the prefix is explained in the book of Fedden's life, *By Jupiter* by Bill Gunston. The 'F' stood for Fedden and 'B' for Butler, one of Fedden's early associates in the days before Fedden joined the Bristol Aeroplane Company. The first blade drawing (made from magnesium alloy) for the Gloster Gauntlet propeller and fitted to a Bristol Mercury engine, carried the number FB593919 and the General Arrangement drawing for the complete airscrew was FB42548. The earliest Rotol magnesium blade drawing is RA.131, a copy of which can still be found in the microfilm system. Other Rotol drawing number systems during the period, using different prefixes, are far too numerous to detail here but as they only involved schematic or other drawings, they can be ignored.

The RA and G series continued for the period up to and beyond the time of the take-over by the Dowty Group, when a completely new system was introduced by Mr S.C. Ilett, the Dowty (Arle Court) Chief Draughtsman. This was based on a nine-digit number and not long after it began to be used it was altered slightly and has continued in this form up to the present day, although after some thirty years it may well have to be modified in the not too distant future as its breakdown of numbers limits the quantity available in each category.

The Rotol Type Number System was intended to be the reference for the complete piece of assembled equipment, whether it was propeller, gearbox, drive shaft, fuel flow proportioner etc and any Rotol multi-part product carried a separate Type Number. On the early accessory gearboxes, that fitted to the starboard engine of the Vickers Wellington Mk IV. carried the number G.083 for the general arrangement drawing with the Equipment Type Number AG.1/1; the port engine gearbox was G.084, Type AG.1/2. The Rotol Type Number System is not straightforward but as it may be of interest to historians, I have made an attempt to explain its workings so far as propellers are concerned.

Extract from first page Rotol Type Number Register – Propellers:

Type No	Aircraft	Engine	G.A.	Blade
AL.1	Gauntlet II	Mercury VI	FB42548	FB593919
AL.2	Vildebeest	Perseus VIII	RA806	FB54246
AL.3	Bombay	Pegasus	RA128	RA132
(These three propellers were only adjustable on the ground)				
No 4 size internal cylinder propeller with 16° range				
4T/RH	Blenheim	Mercury	RA392	RA10059
R4T/1	Blenheim	Mercury	RA1413	
R4T/2	Blenheim	Mercury	RA9453	RA10059
No 4 size internal cylinder propeller with 20° range				
R4/1	Master I	Kestrel	RA928	DB230/C
R4/2	Master I	Kestrel	RA3249	RA613
R4/3	Master II	Mercury	RA4122	DR705
R4/4	Master I	Kestrel	RA8346	DR230
R4/5	Sea Otter I	Mercury	RA9913	DS371
No 5 size internal cylinder propeller with 20° range				
R5/1	Whitley IV	Merlin IV	RA570	RA4042
R5/2	Whitley V	Merlin X	RA570	RA4042
R5/3	Fulmar I	Merlin VIII	RA1571	DB191A
R5/4	Albacore	Taurus		
R5/6	Fulmar II	Merlin XXX	RA1571	DB191A
No 5 non-standardised external cylinder with 35° range				
RX5/1	Spitfire II	Merlin XII	RA3127	RA640
RX5/2	Hurricane I	Merlin III	RA3128	DR291
RX5/3	Spitfire II	Merlin XII	RA4608	RA675
RX5/4	Spitfire I	Merlin III	RA3127	RA640
RX5/5	Hurricane I	Merlin III	RA5653	RA4067
RX5/6	Whitley V	Merlin X	RA8610	RA4060(Dural)
RX5/7	Fulmar II	Merlin XXX	RA8799	DR531
RX5/8	Fulmar II	Merlin XXX	RA9663	RA4065
RX5/9	Fulmar I	Merlin VIII	RA8799	DR531
RX5/10	Spitfire V	Merlin 45	RA9186	RA690
RX5/11	Barracuda I	Merlin 30	RA10534	DS940
RX5/12	Hurricane II	Merlin XX	RA10534	DS940

(Blades with a D prefix eg DS940, DR291 were Weybridge type)

The system was extended with further prefixes: RE – Rotol electric propeller; RXF5 – As RX5 but feathering airscrew; RS5 – Standardised airscrew (capable of increasing the range from 35° to 65° by removing a distance piece from behind the piston).

Thus, the propeller Type Number for the Hurricane Mk IIc was RS5/23 (the end digits were merely the next number in the book) and the propeller Type Number for the Lancaster II (with Hercules radial engines) was RE9/36/4 (in this case the 9 did not indicate shaft size and the '4' related to a 'family' of props).

At a later date the propeller Type Number system was changed in an attempt to simplify matters, using suffixes and starting with R1 in numerical progression: R19/5F5/1 was the propeller fitted to the Spitfire Mk XIV. (5 blades, 'F' fighter, No 5 SBAC Standard shaft size.)

This system was amplified at a later date to allow for more variations in a given Propeller Type: R130/4–20-4/12 was the Propeller Type for the Viscount 802. (R130 – the next number in the Type register book, 4 the number of blades, 20 the propeller root size, the latter number 4 was the SBAC Standard shaft size and the final 12 indicated that it was number 12 in a range or family. For comparison, the Viscount 700 version was R124/4–20-4/10.)

Appendix 2: Prices of Early Airscrews

The following figures are quoted from the Board of Directors' Meeting 18th April 1940, relating to prices agreed with the Air Ministry:

Contract 683671/37 Mercury Blenheim
65 Airscrews £808. 0. 0.
65 Constant Speed Units 42.15. 0.

Contract 751864/38 Merlin Whitley
756 Airscrews £903. 0. 0.
756 Constant Speed Units 39. 5. 0.

Contract 958497/39 Kestrel Master
550 Airscrews £1,275. 0. 0. (for 2 only)
600. 0. 0. (remaining 548)
538 Constant Speed Units 63. 0. 0. (for 2 only)
42.10. 0. (remaining 536)

Appendix 3: Working Hours and Rates of Pay

These two subjects, plus the more general term, working conditions, involving holiday agreements etc have been traditionally negotiated between the company and the trade unions concerned. The first trade union for the skilled workers on the shop floor at Rotol was the Amalgamated Engineers Union, although employees also belonged to other smaller unions such as the Sheet Metal Workers Union and the Metal Mechanics Union. With the influx of women on the shop floor, the AEU tended to ignore them as being unskilled labour and thus it was that a second union, the Transport and General Workers Union, stepped in and accepted membership from the majority of the work-force who came into this category. There is thought to have been only one staff union at that time, the Association of Engineering and Shipbuilding Draughtsmen (AESD) but they were often hampered by lack of numbers in the early days.

No records have been made available to the author concerning negotiations at Rotol, but we may assume that most changes that took place concerning the work-force came about because of discussions between management and unions. Some changes resulted from negotiations at district and national level, with only minor input from the factory level.

Rates of Pay

It has not proved possible to give general pay figures for either staff or works personnel prior to 1959 with any degree of accuracy, although they may still exist in the archives. From discussion with men on the shop floor it is known that a complicated bonus system operated, based on payment by results, but this was always a bone of contention and gave rise to much bitter argument.

When Rotol was formed, certain key individuals were offered service agreements, defining their salary scale and conditions of service. This was applicable to a number of staff and works personnel, both as an inducement to obtain suitable personnel and also a safeguard to ensure they did not leave the company's employment too readily. At the same time it guaranteed employment for a minimum period of three years, and was generally followed by a further agreement and an increase in salary.

Senior management personnel participated in a profit sharing scheme, ranging from an extra £100 for those over £750 per annum to £35 for those receiving less than £400. The bonus to be paid for every £25,000 of profit per annum.

At one period, skilled personnel from the Experimental Shop were given an annual bonus, because they were regarded as key figures and of special importance to the firm. In 1939, the following six people participated in the inauguration of the scheme: R.J. Smith (Turner), H. Jenner (Fitter), C. Beasant (Miller), A. March (Wood Blade Maker), W. Templeton (Inspector), A. Restall (Charge Hand) but other names were added in the years that followed and I am assured that although the bonus began at £25 it was doubled in later years to £50.

Overtime payments for weekly staff were begun in either February or March 1939 after discussions with the Directors, at which it was noted that Rolls-Royce paid for overtime but BAC did not, whereas both Gloster Aircraft Company and Aircraft Components (the forerunner of Dowty Equipment) did. It was recognised that the company could not afford to lose the services of its employees, (particularly of draughtsmen, who were always in short supply) merely because of an overtime inducement from a company such a short distance away.

The following details of working hours and rates of pay have been extracted from *Ladies may now leave their machines* by Diana Murray Hill, which has already been mentioned in the chapter on sources, as a useful aid to understanding what it was like to work at Rotol during the period from 1942 to 1944, and are included here for interest:

Dayshift – Start 8.30 am. Finish 7.30 pm. Lunch hour from 1.00 pm to 2.00 pm Tea break in canteen from 4.15 pm to 4.30 pm (with a brief stop of 10 minutes in the morning about 10.00 am for a cup of tea from the trolley).

Nightshift – This was based on a 'fortnight about' (two weeks on days, followed by two weeks on nights). Starting at 7.30 pm and finishing at 6.30 am. The dinner break was from 12.30 pm to 1.30 am.

Rates of pay – D.M. Hill gives a figure for a woman capstan-operator of 22/- plus 26/- National Award (flat-rate) for a 50 hour week. This applied until she graduated to 'the usual piecework bonus rates, roughly 6d. for every hour you "make" over flat-rate'. Her weekly pay varied from as little as 35/- to an occasional £5. The deductions she quotes are 1/7 for National Health/ Unemployment and 11d. for overall cleaning. At one period she worked every other Sunday, which was paid at Double Time.

Although there is every likelihood that the figures quoted above are correct, the author wishes to point out that they may not be in accordance with ex-employees memories, and in certain instances conflict with the text of the present book.

Apprentice Wage Rates

These figures are given in a document dated 1952: 1st Year age 16 – £2.14.5, rising to 5th Year age 21 – £6. 8.4

Working Hours and Rates of Pay

The following figures for hours of work and rates of pay are the only official ones that have been made available from a company document dated March 1959 and are briefly set out below:

Works and works staff – 5 day 44 hour week.
Start 7.57 am. Finish 5.30 pm.
Lunch break 12.30 pm to 1.15 pm.
Break (works only) 10.00 am to 10.10 am.

Staff (tech and admin) – 5 day 38 hour week.
Start 8.54 am. Finish 5.30 pm.
Lunch break 12.30 pm to 1.30 pm.

Works – nightshift.
4 shifts – 44 hour week. Monday to Thursday.
Start 8.00 pm. Finish 7.45 am.
'Lunch' break 1.30 am to 2.15 am.
Break (1) 10.00 pm to 10.10 pm.
Break (2) 4.15 am to 4.25 am.

Rates of Pay 1959

The rates of pay for manual workers are somewhat too complicated to go into in great detail, and I have attempted to summarise them. They vary between the Rotol departments; Inspection, Tool Room, Stores, Labourers, Works Police, Apprentices etc, etc and of course are graded according to skill – Senior Foreman, Leading Hand, Setter, Progress Chaser, Driver – the list is almost endless, each with a different rate.

The average earnings for a 44 hour week range from £18.2.6 for a senior foreman to £8.12.0 for a labourer. As these rates of pay were negotiated on an annual basis, one can only express sympathy with the persons responsible on both the management and union side for such a valiant effort, year after year.

Overtime payments were also laid down; Works personnel were paid at time and one third beyond the first two hours of normal shift and time and a half thereafter. Saturday was paid at time and a half and Sunday at double time.

Staff rates of pay were only marginally less arduous for the negotiators – of which the author was one – as the staff at this period came under four different trade unions. Draughtsmen over 25 received £13.10.0 (With extra £1 for HNC) while clerical staff rates started at £7.1.6 for a female. These figures have excluded rates for juniors and traineees.

Staff rates of overtime were paid at a flat rate from Monday to Friday and time and a third for weekend.

Appendix 5: Wood Blade Markings

ROTOL LIMITED, GLOUCESTER.
SPECIFICATION No. RAS.150. ISSUE 1. PAGE 6.

IDENTIFICATION MARKINGS ON CAMBER FACE OF WOOD BLADES.

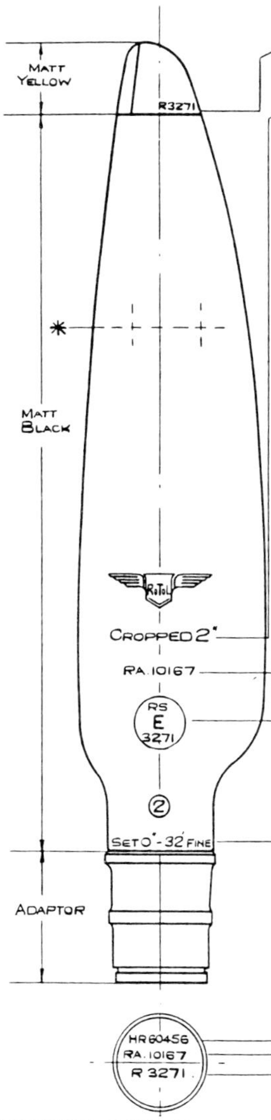

① **SERIAL NO.**
BLACK FIGURES 1/4" HIGH ON YELLOW TIP TO FACILITATE IDENTIFICATION IN STORES.

② **CROPPED.** (IF APPLICABLE)
YELLOW FIGURES 1/2" HIGH, INCLUDING LENGTH OF CROPPING.

③ **DRAWING NO.**
YELLOW FIGURES 1/2" HIGH.

④ **IDENTITY DISC.**
COLOURED DISC, 2 1/2" DIA, WITH THE FOLLOWING INFORMATION:-

(a) **COLOUR.**

PINK INDICATES ROTOL WOOD.
YELLOW " SPRUCE OR DOUGLAS FIR.

(b) **COVERING AND SHEATH.**
INDICATED BY SMALL LETTERS, THUS:-

	RS	R	RS	SS	
MAT'L	ROTOL WOOD.	ROTOL WOOD.	DOUGLAS FIR.	DOUGLAS FIR.	
COVER	ROTOLOID.	ROTOLOID	RAYOID	SCHWARZ	
SHEATH	SIMPLE.	NONE.	SIMPLE.	SIMPLE.	

(c) **BALANCE.**

1 - TO A MASTER MOMENT.

RS M 1496 — INDICATED BY A SINGLE LETTER 1" HIGH. BLADES WITH SIMILAR DISC MARKINGS ARE INTERCHANGEABLE PROVIDED THE COMPLETE PROPELLER IS BALANCED. IN EMERGENCY SUCH A BLADE MAY BE REPLACED PROVIDED THAT SMOOTH GROUND RUNNING IS OBTAINED AND THAT THE COMPLETE PROPELLER IS BALANCED AS SOON AS POSSIBLE.

2 - IN SETS.

RS 4525 4526 4527 4528 — THE LARGE No INDICATES THE BLADE BEING HANDLED. OTHER BLADES IN SET ARE INDICATED BY SMALLER FIGURES.

⑤ **INTEGRATION CORRECTION.**
INDICATES ANY DIFFERENCE FROM NORMAL SETTING REQUIRED FOR AERODYNAMIC BALANCE.
MEASURED AT 0·7 RADIUS POSITION.

* 0·7 RADIUS INDICATED, AS SHOWN DOTTED, ON REVERSE (FLAT) SIDE OF BLADE.

⑥ **STAMPED ON BLADE ROOT BASE.**

1. BLOCK No
2. DRAWING No
3. SERIAL No

Appendix 4: Wood Blade Construction

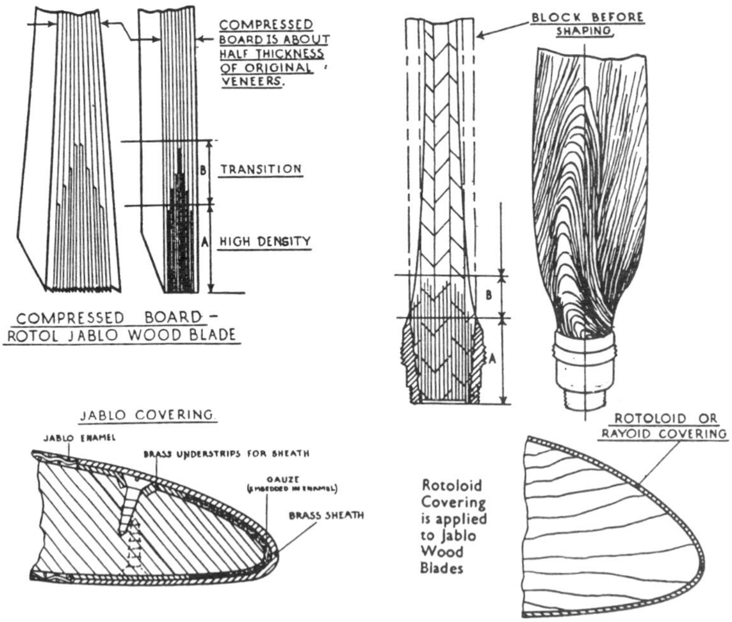

Construction of Rotol Jablo Wood Blade

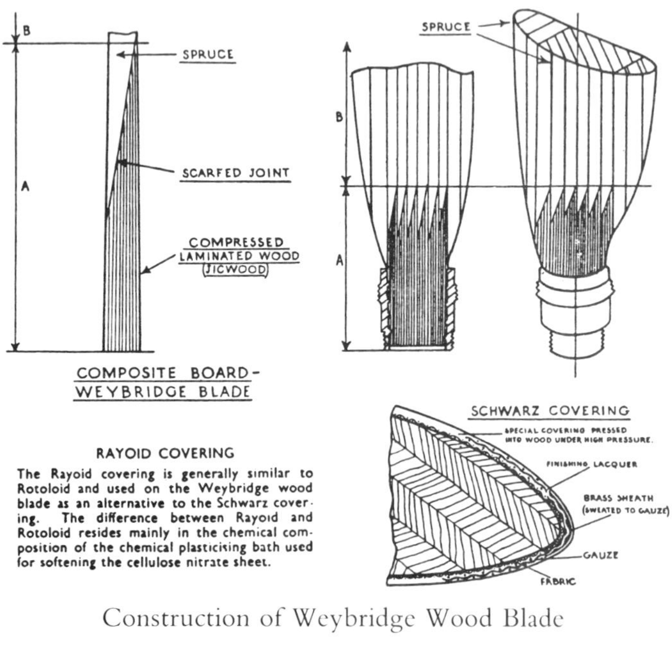

Construction of Weybridge Wood Blade

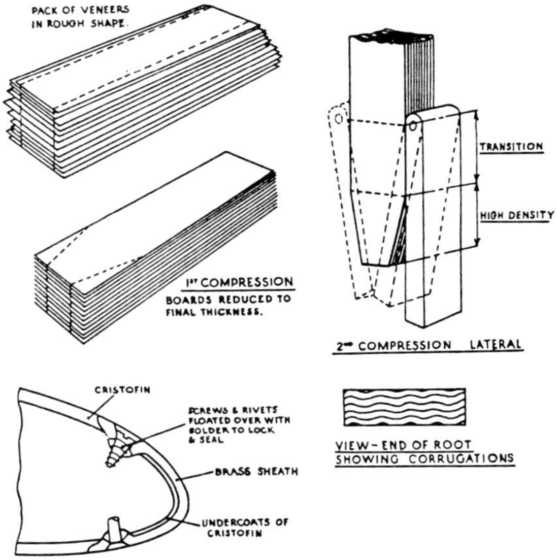

Construction of Hydulignum Wood Blade

Appendix 6: ARP Organisation

AIR RAID PRECAUTIONS ORGANISATION

CONTROLLER :—A. Wilkinson. Telematic No. 255

CHIEF OFFICER :—B. M. Southwell. Telematic No. 190

Mr. Southwell is responsible for the administrative side of A.R.P. Three Deputies cover the full 24 hour day period for emergency at Operations Control and can be contacted on Tel. 190.

Alarm Control Staff are responsible for Factory Evacuation and can be contacted on Tel. 222, but whenever possible, enquiries on A.R.P. matters should be made to Mr. Southwell on Tel. 190.

VOLUNTEER ORGANISATION—SECTION LEADERS.

DAY SHIFT.		NIGHT SHIFT.	
Wardens' Service		**Wardens' Service**	
A. A. Smith	tel. 279	C. Grinnell	tel. 145
C. Gooderman	tel. 145	W. Tombs	tel. 145
G. Bartrip	tel. 145	F. Attwood	tel. 196
J. Johnston	tel. 198		
Casualty Service and F.A.P.		**Casualty Service and F.A.P.**	
V. Dent	tel. 178	Mr. Reay	tel. 178
R. Paish	tel. 170 or 168		
R. Wheeler	tel. 172		
Rescue Service		**Rescue Service**	
E. Leah	tel. 215	Mr. Farmer-Little	tel. 105
C. Wacks	tel. 105		
Decontamination		**Decontamination**	
C. Rawcliffe	tel. 163	Mr. White	tel. 105
Aircraft Identification		**Roof Fire-Watcher**	
J. Tallett	tel. 196		tel. 230
A. Fry	tel. 196	**Bomb Disposal Unit**	
S. Rea	tel. 269	Mr. Harris	Chelt. 52908 (telc. 251)

Appendix 7: Wartime Transport Time Table

PASSENGER AND MAIL TRANSPORT SERVICE TO DISPERSAL AREAS

GLOUCESTER

ROTOL	TECHNICAL COLLEGE		REPAIR DEPT.		ACC'S REPAIR DEPT.		ROTOL
DEPART	ARRIVE	DEPART	ARRIVE	DEPART	ARRIVE	DEPART	ARRIVE
8.45 A.M.	9.00 A.M.	9.40 A.M.	9.05 A.M.	9.35 A.M.	9.15 A.M.	9.25 A.M.	9.55 A.M.
10.05 A.M.	10.20 A.M.	11.00 A.M.	10.25 A.M.	10.55 A.M.	10.35 A.M.	10.45 A.M.	11.15 A.M.
11.25 A.M.	11.40 A.M.*	12.20 P.M.	11.45 A.M.	12.15 P.M.	11.55 A.M.	12.05 P.M.	12.35 P.M.
2.00 P.M.	2.15 P.M.	2.55 P.M.	2.20 P.M.	2.50 P.M.	2.30 P.M.	2.40 P.M.	3.10 P.M.
3.20 P.M.	3.35 P.M.	4.15 P.M.	3.40 P.M.	4.10 P.M.	3.50 P.M.	4.00 P.M.	4.30 P.M.
4.40 P.M.	4.55 P.M.	5.35 P.M.	5.00 P.M.	5.30 P.M.	5.10 P.M.	5.20 P.M.	5.50 P.M.

* DELIVERY OF MAIL TO GEORGE ST. POST OFFICE.
LAST JOURNEY THURSDAY WILL START 5 MINS. BEFORE TIME.
LAST JOURNEY SATURDAY MORNING WILL START 5 MINS. BEFORE TIME.

PASSENGER AND MAIL TRANSPORT SERVICE TO DISPERSAL AREAS

CHELTENHAM

ROTOL	IMPERIAL HOUSE		SPINNER FACTORY		KEYNSHAM BANK		ROTOL
DEPART	ARRIVE	DEPART	ARRIVE	DEPART	ARRIVE	DEPART	ARRIVE
8.45 A.M.	9.00 A.M.	9.35 A.M.	9.05 A.M.	9.30 A.M.	9.10 A.M.	9.20 A.M.	9.50 A.M.
10.00 A.M.	10.15 A.M.	10.30 A.M.	10.20 A.M.	10.45 A.M.	10.25 A.M.	10.35 A.M.	11.05 A.M.
11.15 A.M.	11.30 A.M.	12.05 P.M.	11.35 A.M.	12.00 NOON	11.40 A.M.	11.50 A.M.	12.20 P.M.
2.00 P.M.	2.15 P.M.	2.50 P.M.	2.20 P.M.	2.45 P.M.	2.25 P.M.	2.35 P.M.	3.05 P.M.
3.15 P.M.	3.30 P.M.	4.05 P.M.	3.35 P.M.	4.00 P.M.	3.40 P.M.	3.50 P.M.	4.20 P.M.
4.30 P.M.	4.45 P.M.	5.20 P.M.	4.50 P.M.	5.15 P.M.	4.55 P.M.	5.05 P.M.	5.35 P.M.

Appendix 8: Propeller Design

Readers may be interested to have a few basic details of the way modern propellers are designed and I am grateful to Mrs C. Wain, a long-serving member of the Dowty Rotol Structures Department for help with this section:

> Information is received from airframe manufacturer and/or engine manufacturer with regard to size of aircraft, flight envelope and basic engine data. Frequently the aircraft maker will specify propeller diameter, with due regard to the length of undercarriage which of course limits the maximum size, and the total number of blades.
>
> From this data, performance and noise level engineers will design the blade shape, which is usually a compromise between performance and noise requirements. The data provided would include (a) blade chord (b) blade twist (c) blade thickness and (d) blade camber, to suit a particular section (ARA-D, NACA 16, CLARK Y etc.).
>
> Structural designers use this data to calculate stress and frequency analysis of the blade. This work could then produce a final blade design, that would again be a compromise from the original performance/noise data ie a blade that is ideal for performance requirements would not necessarily give satisfactory stress loads or frequency characteristics.
>
> The stress analysis also generates blade loads which are then used in the design of the propeller equipment ie Hub, bearing, pitch changing mechanism. These blade loads include centrifugal force (CF), static and vibratory bending moments (BM), and centrifugal twisting moments (CTM). CF and BM's are used in the hub and bearing design, with CTM and aerodynamic twisting moment (from performance analysis) being used in the pitch control system.

This has obviously simplified an extremely complex subject, if the reader wishes to delve deeper into it, there are several text books and papers available.

Roger Armstrong was kind enough to give some background details to the introduction of computers to the company, which has made such an enormous contribution to the design of propellers:

> During the war, computing aids were confine to copies of Chambers 7-figure tables, 10″ slide rules, Fowlers 500″ cylindrical slide rule and two Brunsviga and one Marchant hand calculating machines. At the peak, about 30 girls operated with the Fowlers slide rules and their work consisted mainly of propeller performance and stress calculations. The official requirement for these girls was to be good at figures but in view of the fact that at one time no less than three beauty queens were resident in the department, it was said that they were recruited on their personal figures!
>
> In the early 1950's a problem with propeller vibration was solved by the use of 20″ × 20″ matrices and the girls gallantly tackled the job, but any error was soon spread across the whole matrix and tedious checking was necessary, and it took two girls 21 days to complete one frequency. The first solution was the use of punch-card equipment from the British Tabulating Company, using the 506 Multiplier. This was an electro-mechanical machine with hundreds of relays, and was very unreliable, so much so that a second machine had to be used and the answers compared stage-by-stage. Each stage consisted of merging 20 cards within two packs of 400 cards and then segregating the 400 cards and comparing these with the 400 cards from the second machine. The whole process took fifteen days and

produced four blade frequencies – assuming that no mistakes had been made.

These electro-mechanical machines worked in the denary system but one had been converted to work in pounds, shillings and pence which could cause problems for the unwary. Both machines were replaced by a 542 electronic calculator which worked in coded denary. Breakdowns of the electronics were unknown but the card reading was done by wire brushes making contact through holes in the card, which caused problems. Using the 542 upped the speed of the calculations to eight cases in five days.

The next step was the acquisition of an Elliott 402E computer which was housed in the brick technical block, requiring its own generator and air conditioner – to keep the whole thing cool – and was 14ft long, 18in wide and 6ft high. It was programmed in machine code and had drum storage with 128 words per track, 6 tracks always available and the 7th could be selected from 31 further tracks and was available one revolution after the switching order. Each word on the track was therefore available once every drum revolution except for 16 words which were immediate access, hence optimum programming was essential to achieve the best utilisation. Data input and output was by 5-hole paper tape, and the contents (orders or numbers) of the machine could only be inspected as 32 bit binary numbers. It worked as a fixed part machine and was truly a first generation computer.

Surprisingly, despite the fact that the machine was entirely a valve machine, its serviceability and reliability was very good. This could only be attributed to the meticulous care given to it by Martin Wright, the computer maintenance engineer. In those days the programmers were inclined to blame the machine when their programmes did not work and Martin in turn blamed the programmers – and he was generally right! (No pun intended). While on the personaltities in the department it is worth recording that on one occasion when a technician was appointed computer programme engineer, a colleague was heard to remark, 'Engineer! I wouldn't make him second engineer on a concrete mixer.' Beryl Price was the Elliott operator and could happily sit and knit while it clicked away, occasionally getting up to put the case it was working on back into the machine. This was because she could detect the occasional sound sequence of track switching had changed, when a relay was slow.

The installation cost in 1951 was £35,000 and Rotol had the fifth and last model 402E that was built. The vibration cases referred to earlier, which took 15 days for four cases on the electro-mechanical multiplier, now took 35 minutes per case on the Elliott, leaving ample capacity for propeller blade stressing and performance calculations. This enabled blade designs to be 'massaged' to give optimum design and finally took guessing and intuition out of blade design.

The SBAC standard propeller performance estimation method involved a series of relatively simple calculations with data derived from various curves and in particular families of curves. Unfortunately these curves had originally been faired by hand and the girls interpolated by eye. The not unreasonable demand was made that answers from the Elliott must agree with the hand calculations. This caused major problems with interpolation methods used but they were eventually solved and the answers agreed.

The strip propeller performance method using ten representative stations along the blade required lift and drag coefficient against angle of attack and Mach number. The basic aerofoil section data was stored on master tape and consisted of 256 track of data with C_L and CD in the same word. ($-C_D$ meant that the data was extrapolated). Because of storage limitations of the Elliott this master paper tape was read by a special programme and reduced by interpolation to 20 tracks appropriate for the thickness/chord ratio and section camber ie 2 tracks for each station. These 20 tracks were stored on paper tape and could be used anytime the aerodynamic properties of the blade were required for performance or aerodynamic loads required by the structural programmers.

Hence the computer could supply everything the aircraft manufacturer required and also all the loads required for structural analysis of the blade by the computer and the loads imposed by the blade on the hub.

Finally the time history programme of the Dart propeller control proved to be very valuable but its input data was so difficult that a separate programme was written to check it. This was a wonderful example of a digital computer doing a job which was really ideal for an analogue computer.